T0247645

The Watch's Wild Cry

A VOYAGE ABOARD THE WHALING VESSEL *CLARA BELL*

ROBERT F. WEIR

EDITED BY
ANDREW W. GERMAN

Essex, Connecticut

An imprint of The Globe Pequot Publishing Group, Inc.
64 South Main Street
Essex, CT 06426
www.globepequot.com

Distributed by NATIONAL BOOK NETWORK

British Library Cataloguing in Publication Information available

Library of Congress Cataloging-in-Publication Data

Names: Weir, Robert F., 1836–1905, author. | German, Andrew W., 1950– editor.
Title: The watch's wild cry : a voyage aboard the whaling vessel Clara Bell / Robert F. Weir ; edited by Andrew W. German.
Other titles: Voyage aboard the whaling vessel Clara Bell
Description: Essex, Connecticut : Lyons Press ; Mystic Seaport Museum, [2024] | Includes bibliographical references.
Identifiers: LCCN 2024030377 (print) | LCCN 2024030378 (ebook) | ISBN 9781493081042 (cloth) | ISBN 9781493081059 (ebook)
Subjects: LCSH: Weir, Robert F., 1836–1905—Diaries. | Weir, Robert F., 1836–1905—Travel. | Clara Bell (Bark) | Whaling—Massachusetts—New England—History—19th Century.
Classification: LCC G545 .W447 2024 (print) | LCC G545 (ebook) | DDC 639.2/8092 [B]—dc23/eng/20240711
LC record available at https://lccn.loc.gov/2024030377
LC ebook record available at https://lccn.loc.gov/2024030378

∞™ The paper used in this publication meets the minimum requirements of American National Standard for Information Sciences—Permanence of Paper for Printed Library Materials, ANSI/NISO Z39.48-1992.

Mystic Seaport Museum is the nation's leading maritime museum. Founded in 1929 to gather and preserve the rapidly disappearing artifacts of America's seafaring past, the Museum has grown to become a national center for research and education with the mission to "inspire an enduring connection to the American maritime experience."

Contents

Preface

When Robert F. Weir ran off to sea at age 19, he knew he had some of the Weir family's proficiency at art, so it is not surprising that he recorded his experiences in both words and images. For the rest of his life, Weir clearly valued the journal he kept and then illustrated to document his introduction to whaling life on board the bark *Clara Bell* from August 1855 to May 1858. Even 40 years after completing it, he was still revisiting it and making notes.

After Weir's death in 1905, the journal likely remained in his wife Anna's possession until her death in 1910. Since they had no children, it is not clear what hands it passed through before a resident of Mystic, Connecticut, donated it to the Marine Historical Association, now Mystic Seaport Museum, in 1943. It became Log 174 in the Museum's collection, and ever since, it has been the Museum's premier example of an illustrated whaling journal—as well as an engaging account of a young man's maturation in the dangerous, exciting, exotic, tedious, and often dreadful world of the whale hunt.

For more than 80 years the Museum's curators and librarians have been diligent both to protect the volume from damage and to make it accessible to users. From Donald Judge to Douglas Stein to Paul O'Pecko, they have properly housed it, had the images photographed for use, photocopied the journal's pages to prevent wear to the original, and, finally, digitized it for online access.

At the suggestion of Paul O'Pecko, this published version continues the effort to make the journal more broadly accessible. To make it intelligible for current and nonmaritime readers, we have edited the text lightly, spelling out abbreviations and numerals and regulating punctuation and capitalization. As part of that effort, I added contextual notes to elaborate on vessels, places, things, or processes mentioned by Weir. To provide context on Weir himself, I have added introductory and postscript details on the Weir family and Robert F. Weir's life, as much as we can reconstruct it. Because he went on to serve in the US Navy during the Civil War, we have added samples of sketches he sent home and others that were published in *Harper's Weekly*.

Besides Paul O'Pecko, many individuals deserve thanks for their contributions to the project. Mystic Seaport Museum curator of maritime history Michael P. Dyer, formerly of the New Bedford Whaling Museum, added information and provided a critical review of the notes, introductory text, and postscript. Maribeth Quinlan, the Museum's collections access and research manager, promptly answered many questions and assisted with a variety of materials in the collection. Claudia Triggs, the Museum's intellectual property coordinator, managed the outreach to other institutions and delivery of images for the book. Mary Anne Stets oversaw the entire process for the Museum.

At the Mariners' Museum and Park, Lisa M. Williams assisted with the request for an image, while Sharon Kidd assisted at the Pearce Museum at Navarro College, Emma Rocha provided an image from the New Bedford Whaling Museum, and Donna E. Russo, library and archive specialist at Historic New England, made available a photograph of Weir's favorite sister, Emma Weir Casey.

For Lyons Press, acquisitions editor Brittany Stoner helped develop the concept and managed the project.

Just as a vessel requires a strong crew for a successful voyage, a talented crew brought Robert F. Weir's *Clara Bell* voyage to publication. Thanks to all of them.

Introduction

In 1855, Robert Fulton Weir ran off to sea, feeling ashamed for disappointing his family. It was a very large and prominent family to disappoint, especially the patriarch, Robert Walter Weir.

Robert Walter Weir was the son of Robert Weir, who had been born in 1770 at Paisley, Renfrewshire, near Glasgow and the River Clyde in western Scotland. Robert was the third of about 14 children. As a young man, he emigrated to New York as a merchant, and there he met Maria Catherine Brinckley. Robert was naturalized on August 17, 1802, and 10 days later, he and Maria wed. Born in June 1803, their son Robert received his grandfather Walter's name as a middle name. He was the oldest of seven children.

Successful during the years of neutral trade, Robert Weir lost much of his fortune when shipping was curtailed during the War of 1812. Robert W. Weir initially followed his father into the shipping business, but at age 18, he gave up a clerkship to become an artist. Essentially self-taught, he trained in New York for three years before traveling to Italy in 1824. From 1824 to 1827, he studied the works of the Italian Renaissance masters in Florence and Rome before returning to New York to provide care for a sick friend. In 1829, he was elected to the National Academy of Design, an honorary association of American artists that had been founded by a group of prominent young artists in 1825. Weir developed himself as an occasional portraitist and landscape painter but principally a painter of historical scenes, somewhat in the tradition of the Italian artists he had studied.

In 1829, Robert W. Weir married 23-year-old Louisa Ferguson, daughter of John Ferguson, who had served as mayor of New York City in 1815. The Weirs lived in New York City as he developed his reputation, and their first children, Walter (1831–1898) and Louisa (1832–1919), were born there.

In 1834, Robert W. Weir was appointed to the position of drawing instructor at the US Military Academy, about 50 miles up the Hudson

River at West Point. The family moved into faculty housing there, and the children would grow up around many of the most promising military figures of the period. Because accurate sketching and cartography were considered essential military skills, Weir instructed each class of cadets in figure drawing, landscape, cartography, and aerial perspective. He also continued to accept artistic commissions. In 1837, in its effort to decorate the US Capitol with views of the origins of the United States, the US Congress commissioned Weir to paint an imagined view of the departure of the Pilgrims for North America in 1620. Completed and hung in the Capitol Rotunda in 1843, *Embarkation of the Pilgrims* measures 12 by 18 feet.

Robert Walter Weir (1803–1889), ca. 1864. (Smithsonian American Art Museum, Washington, DC)

At West Point, Weir's family grew in number, much like those of his father and grandfather. The children born at West Point were Emma (1834–1911); Robert Fulton (January 1836–1905), who received the surname of Robert's Scottish grandmother as a middle name; Gulian Verplanck (December 1836–1886); Henry Cary (1839–1927); John Ferguson (1841–1926); Mary (1843–1848); and Alice (1845–1845). Their mother, Louisa, died from complications of Alice's birth in January 1845, and Alice died that August. Suddenly, Robert had to manage a family of eight, between the ages of 14 and two. To serve as a housekeeper, 29-year-old Susan Martha Bayard joined the household. She was the daughter of the New York Episcopal clergyman Lewis Pintard Bayard, who had died in 1840, leaving the family impoverished. In July 1846, Robert married Susan. As the children learned to live with their stepmother, the second Weir family expanded as well, including Bayard (1847–1879),

William Bayard (1849–1879), Charles Gouveneur (1850–1935), Julian Alden (1852–1919), Anne (1853–1932), Carrie (1855–1937), and Helen (1857–1939).[1]

The family lived in the northernmost of the stone duplex quarters on Professors' Row at the Military Academy, with a view up the Hudson. Robert F. Weir would retain fond memories of attending the Academy's small chapel, built in the year of his birth, with his father's mural, *Peace and War*, behind the altar. Since the chapel usually functioned as a Presbyterian church, in 1841 his father designed a stone Episcopal church for the village of Highland Falls, just south of the Military Academy, where the laborers who served the Academy lived. Partly funded by Weir's earnings from his painting, *Embarkation of the Pilgrims*, this church was completed and dedicated in 1847 as the Church of the Holy Innocents. The family then attended services there and the children received Episcopal religious instruction. During the week, they would gather in their father's studio for morning and evening prayers. Living in a devout household under a military schedule, the Weir children faced high expectations. At the same time, John F. Weir would recall: "At Home I was always taught to look upon the gloomy side of life. Father has his ideal, which causes him to be borne along with different tides withersoever they tend, taking an unsuspecting, confiding view of human nature."[2]

Robert W. Weir found a protégé in Cadet Truman Seymour, a member of the West Point Class of 1846, many of whom would serve with distinction during the Mexican–American War and then the Civil War. Seymour would return to West Point as Weir's assistant professor of drawing from 1850 to 1853 and would marry Weir's eldest daughter, Louisa, in 1852. During his time at West Point, although Seymour was 12 years older, he seems to have become close to young Robert F., likely through a shared interest in drawing.

Growing up, Robert F. was especially close to his sister Emma, who was two years older. By 1850, Walter had gone off to school in Hartford, Connecticut, where he would become a teacher and marry in 1855. After Louisa's marriage, Emma would marry Thomas Lincoln Casey, who graduated first in his class at West Point in 1852. He was serving as assistant professor of engineering there when they married on May 12, 1856.

Robert W. Weir's studio was an engaging place, with a large, heavily carved Dutch chest, a suit of armor, plaster casts of classical sculpture, and souvenirs Weir had brought back from Italy. John F. Weir found the "mystic halo" of thought inspired by the studio pushed him to become an artist even before he knew if he could draw. While Robert F. showed artistic skill and may have received informal instruction from his father, he was not directed to become an artist. Nor did his proficiency on the violin ready him for a musical career. Most of the sons were sent away to school, and Robert may have spent time at some Hudson River academy, but his skills seemed more practical than academic. In the early 1850s, he was placed in a training position at the West Point Foundry.[3]

An Artist's Studio, by Robert's brother John Ferguson Weir, depicts their father Robert W. Weir in his studio at West Point, where the family gathered for daily prayers and young Robert may have received an introduction to drawing from his father. Oil on canvas, 1864. (Los Angeles County Museum of Art. Gift of Jo Ann and Julian Ganz Jr. [M.86.307] Digital Image © 2009 Museum Associates/LACMA/Art Resource, New York)

Established at Cold Spring, New York, across the Hudson River from the US Military Academy, the West Point Foundry began operation in 1817. With abundant iron ore and hardwood for charcoal in the area, it was envisioned as a producer of military artillery and other ironwork. In 1836, Captain James Parrott, an 1824 graduate of West Point, resigned his commission as inspector of ordnance and became superintendent of the foundry. Parrott made it an innovative enterprise. In addition to army ordnance, it produced cast-iron water pipes for New York City, rolling mills for processing sugarcane in the Caribbean, early steam locomotives, and, in 1843, the iron-hulled revenue cutter *Spencer*. At its height of operation, from the 1850s to the 1870s, the foundry employed 1,000 workers and produced 10,000 tons of cast iron per year. In 1860, Parrott would develop his noted and distinctive rifled cannon, which had a cast-iron barrel with a wrought-iron band around the breech to reinforce it. The West Point

The scale of operations at the West Point Foundry, across the Hudson River in Cold Spring, New York, is suggested in John Ferguson Weir's 1866 painting, *The Gun Foundry*, which shows the casting of a Parrott rifled cannon. Robert Weir's time at the foundry, apparently in a training position, seems to have been brief. (Courtesy of Putnam History Museum, Cold Spring, New York)

Foundry would produce more than a thousand Parrott rifles, in various sizes, and millions of artillery shells during the early 1860s.[4]

Robert F. Weir was probably about 16 when he crossed the Hudson to go to work at the West Point Foundry. His role is unknown, although he referred to it as a "course," so he seems to have been part of a training program. With his family background and artistic ability, he might have been an apprentice draftsman, or he may have begun in the foundry's machine shop, steam plant, or pattern-making shop to learn technical skills.

As part of the Hudson Valley elite—a mix of literary, artistic, and financial first-families—at some point, the Weirs had connected with the family of William N. Chadwick, a New York-born industrialist who had settled in Cohoes, north of Albany, in 1841 to develop textile mills there. He served as president of the Harmony Cotton Mills and became a leader of the community. The Chadwicks had six children, the youngest of whom, Anna, was two years older than Robert F. Weir. Like Weir, she was musical and interested in creative writing; she was also outgoing and attracted several young suitors. In the 1850s, she attended a boarding school in New York City, and somewhere between the city and Cohoes, Robert and Anna developed their relationship.[5]

The very large Morris family was near the top of the Hudson Valley elite. Perhaps earlier, but definitely during his time at the foundry, Weir became close to Lewis Morris, son of Richard Rutherford Morris, a wealthy farmer of Pelham, New York, whose grandfather, Lewis Morris, had signed the Declaration of Independence. "Lou" Morris was nine years older than Weir, but at Cold Spring, he was Weir's friend and confidant, encouraging him to "finish my course at the foundry" when he became disillusioned. Lewis Morris died at New York City on March 28, 1855. In his journal, Weir would imply that he felt somehow responsible. Possibly a foolish prank had gone wrong, injuring his friend, or possibly Lewis Morris was ill. Beyond the death of his friend, it has also been suggested that Weir had disgraced himself by gambling.

Whatever heavy weight of guilt he felt, 19-year-old Robert F. Weir left Cold Spring on August 3, 1855, apparently without a clear idea of where to go to hide from his family. The idea of seeking distance and anonymity at sea may have been suggested to him, or perhaps he had read enough to think of seafaring, or whaling, as an escape. Since Richard Henry Dana had interrupted his studies at Harvard to go to sea to restore his eyesight, then published his experiences as *Two Years Before the Mast* in 1840, seafaring had the cachet of adventure and anonymity for privileged young men.

Robert F. Weir felt closest to his sister Emma (1834–1911). This daguerreotype of Emma in her wedding dress was made at the time of her marriage to Thomas Lincoln Casey in 1856. (Courtesy of Historic New England)

For those without seafaring experience but with a thirst for adventure, the whaling industry offered the most opportunity. Europeans had developed the processes for hunting and processing whales at sea by the 1500s. After Nantucket, Massachusetts, mariners took up whaling in the late 1600s, the industry spread as New England merchants and mariners sought new commodities. The insulating blubber of the right whales hunted by Europeans and early New Englanders was rendered into oil for various uses, but Nantucketers expanded the industry and its products when they encountered sperm whales in the open ocean. Sperm whales produced a finer oil, and the waxy spermaceti contained in their forehead structure called the "case" could be molded into clean, bright-burning candles. By the 1760s, when whaling expanded from Nantucket to Bedford on the southeast Massachusetts coast, it had become a characteristic New England maritime industry. After the Revolutionary War, Bedford was rebuilt as New Bedford, and New England whalers pursued the hunt through the South Atlantic, rounding Cape Horn into the Pacific Ocean in the 1790s. After the War of 1812, the demand for whale products increased, partly fueled by the use of sperm whale oil to light the lamps of US lighthouses. By the 1830s, New Bedford had become the leading American whaling port, and the industry itself was perhaps the fifth most valuable American enterprise. Right whale oil was processed into oil for lighting, oil for lubricating machinery, and soap, and sperm whale oil was processed into finer, clean-burning oil for interior lighting and lighthouses and fine oil for lubricating delicate machinery, from watches to the looms of the rapidly expanding New England textile industry. Spermaceti candles remained in high demand, and there was increasing demand for the long strands of pliable baleen from the mouths of filter-feeding right and bowhead whales, which could be manufactured into many flexible items. In 1846, when Robert F. Weir was 10, the American whaling industry reached a peak of 735 vessels. The fleet then began to decline, although the single most profitable year was 1853, and New Bedford's fleet would peak at 329 vessels in 1857.[6]

A fleet of 700 vessels, about half of which sailed from New Bedford, required more than 20,000 men. A whaling crew required a few men skilled in whaling, a few experienced seamen, and 15 to 20 inexperienced "greenhands," who provided muscle and learned their work during the voyage. In earlier decades, whaling crews could be raised locally, but by the 1830s, whaling owners relied on a steady supply of restless teenage boys from interior New England and New York and island men from places where whaleships called, including the Azores in the Atlantic and the many islands of the Pacific.

By the 1850s, there were numerous published accounts of young men who had gone whaling, which might inspire others to follow them. Francis Allyn Olmsted had made a whaling voyage out of New London, Connecticut, to recruit his health after graduating from Yale, and he published his account as *Incidents of a Whaling Voyage* in 1841. Irish emigrant J. Ross Browne had worked on a riverboat before he sailed out of Fairhaven, Massachusetts, in 1842 and published his account as *Etchings of a Whaling Cruise* in 1846. That year, Reuben Delano published his *Wanderings and Adventures of Reuben Delano, Being a Narrative of Twelve Years Life in a Whale Ship*. The best-known work on whaling, by New York native Herman Melville, was the 1851 novel *Moby-Dick, or The Whale*. Melville had sailed aboard the Fairhaven bark *Acushnet* in 1841. In the year that Weir headed for sea, Charles Nordhoff published his account, *Whaling and Fishing.* William B. Whitecar, a young Philadelphian who headed to New Bedford two months before Robert Weir decided to run off to sea, sailed aboard the whaleship *Pacific* and, upon his return, published his account as *Four Years Aboard the Whaleship. Embracing Cruises in the Pacific, Atlantic, Indian, and Antarctic Oceans. In the Years 1855,'6,'7,'8,'9*, in 1860.

Like many before him, Robert F. Weir traveled east, most likely following William Whitecar's route by steamboat from New York to Fall River, and then by train the few miles further to New Bedford. There, like Whitecar, he was likely intercepted by a "landshark" who made his living representing boardinghouses and shipping agents to naïve arrivals. Weir seems to have been directed to the outfitters Barney & Spooner and to a boardinghouse operated by the widow Hope Howland Doane at 23 South Second Street. Boardinghouses like hers lodged a mix of whalemen home from the sea and young men waiting to go to sea. Whitecar found boardinghouse food barely better than that provided at sea, with an emphasis on salt beef and salt pork. While Weir's fellow boarders may have taken advantage of the port's numerous grog shops or even its houses of prostitution, he spent his free time during his two weeks reading. Although he did not mention it, Weir possibly visited the Seamen's Bethel for a worship service, or to use the sailors' reading room there.[7]

Quakers James S. Barney and Charles M. Spooner operated as merchant tailors and outfitters at 25 North Water Street. In addition to selling clothing in the community, they assembled crews for whaleships, paid each man an advance, and provided him with a basic outfit of sailor's clothing, possibly eating utensils, and a sea chest in which to store his possessions. Weir's bill came to $48.71 for his cash advance and outfit, $14.00 for clothing from Henry Harris, and $8.00 for board to Mrs. Doane. The vessel's agent then paid Barney & Spooner, leaving Weir indebted to the vessel for $70.71 before stepping aboard. With several years' interest added, the sum would be deducted from his pay at the end of the voyage.[8]

Whaling crews were not paid a standard wage. Rather, with the length of the voyage and the value of the whale products to be taken both uncertain, they were signed aboard for a set fraction of the proceeds once the cargo was delivered and costs were deducted. Each man's share was called his lay. While a captain might merit a 1/15th lay, mates might receive from 1/20th to 1/55th, and the boatsteerers who harpooned the whales earned about 1/90th, an inexperienced greenhand like Weir would be offered a "long lay" of about 1/195th. Each man's lay was confirmed on the Whalemen's Shipping Paper, or articles of agreement for the voyage. A formal contract, probably signed at the Barney & Spooner store and witnessed by them, this crew list would be carried by the captain on board and amended as men departed or were added to the crew. If a man deserted from the vessel, he forfeited his lay. If he was promoted, his lay was adjusted for the rest of the voyage.

Like many a young man who did not wish to be found by his family, Weir assumed a seafaring identity. Proud of his Scottish ancestry, he chose the surname of the legendary Scottish patriot, Sir William Wallace. As a whaleman, Robert Weir would be Robert Wallace, and he would accept a greenhand's berth on the bark *Clara Bell*, managed by Robert L. Barstow of Mattapoisett, a small whaling and shipbuilding port seven miles east of New Bedford, and commanded by Captain Charles H. Robbins.

Charles H. Robbins was born to Lemuel and Rachael Bailey Robbins at Mattapoisett in 1822. He was the seventh of nine children, only two of whom were males. The family moved to a small, crowded house in the maritime workers' neighborhood of New Bedford in 1828. Three years later, when Charles was nine, his father died, and he began to help support the family, working as an office boy and paper carrier for the *New Bedford Mercury* newspaper. Looking for adventure, and with the prospects for a $50 advance and another $50 upon his return, 15-year-old Charles signed aboard the whaleship *Swift* to serve Captain Lewis Tobey as cabin boy.

Robbins found whaling so engaging, despite Captain Tobey's unpredictable personality at sea, that he taught himself the calculations for navigation and, by the end of the four-and-a-half-year voyage to the Pacific Ocean sperm-whaling grounds, had risen to boatsteerer, harpooning whales and then steering the whaleboat during the killing of the whale. Returning to New Bedford in 1841, he soon sailed on the *Balaena* as a boatsteerer. In 1845, he married Hannah Warren of Charlestown, Massachusetts, and she may have continued to live in the home of her shoemaker father in Boston when Robbins sailed as first mate of the New Bedford bark *Hope* for a sperm-whaling voyage to the Indian Ocean in 1847. After the voyage he returned to Boston, but later that year he took command of the *Hope* for a three-year voyage, again sperm whaling in the Indian Ocean. During his absence, Hannah gave birth to a daughter, who died before her first birthday. Captain Robbins never saw her. In 1855, as he prepared to take command of the whaling bark *Clara Bell*, the Robbinses moved to New Bedford and were living along County Street near his mother.[9]

Launched by Leonard Hammond at Mattapoisett in 1852, the 296-ton whaling bark *Clara Bell* had completed one voyage when Weir joined it. In his book *The Gam*, Captain Robbins later described the *Clara Bell* as a half-clipper, with a sharper bow than traditional whaleships. The hull was painted black, with two narrow white bands from bow to stern, and the figurehead was a "winged-dragon." The vessel was 105 feet long, with a beam (width) of 26 feet. Rigged as a bark, the vessel had square sails on the fore- and mainmast and fore-and-aft sails only on the mizzenmast. This rig had become common for whaling vessels as it offered maneuverability as well as open-sea stability while being more economical than a full-rigged ship.

At the bow was a small raised anchor deck, where the anchors were secured while the vessel was at sea. Just aft of that on the main deck, the vessel had its windlass, or horizontal winch, turned by the crew with a pump break mechanism using long levers to crank it when hoisting anchor and heavy whale parts. Near the windlass was the hatch leading down to the crew's quarters in the forecastle (foc's'le). The small galley or kitchen house seems to have been forward, probably along the larboard rail. Aft of the foremast stood the brick tryworks with two large iron pots for rendering oil from whale blubber, followed by the main hatch leading down to the blubber room between decks. Aft of the mainmast, the booby hatch, with a sliding cover, led down to the steerage where lived the boatsteerers, the carpenter, the cooper, who set up and managed the casks for whale oil, the cook, and the steward, who tended to the officers. Above the booby hatch were the boat skids, a bridge of timbers from rail to rail, well above the deck, on which spare whaleboats were stored upside down.

Aft of the boat skids and mizzenmast, near the companionway to the officers' quarters, was the ship's wheel and compass and a small house on deck. Below the deck, the crew's cramped, often damp bunks were in the fo'c'sle at the bow. Continuing on toward the stern, the large, open blubber room, with low overhead, was used for cutting the large blanket pieces of blubber stripped from whales into pieces small enough to try out (render) in the try pots. Next was the steerage bunkroom. The officers had their own staterooms at the stern: the captain to starboard, with a day cabin across the stern; the first mate in a private stateroom to larboard where he kept the ship's logbook; and the second and third mates in a double stateroom. The officers' staterooms opened on their saloon or dining room, with a table and benches. Off the saloon was a small pantry from which the steward could serve meals to the officers. Below, the hold contained a tightly packed array of wooden casks, large and small, filled with ballast water, drinking water, provisions, and equipment, which would be replaced by casks full of whale oil during the voyage. The standard whale oil barrel contained 31½ gallons, so that became the measure by which a vessel's catch was calculated, although the casks in which the oil was stored varied in size.[10]

Captain Charles H. Robbins (1822–1903), photographed later in life. (Charles H. Robbins, *The Gam*, 1899)

Whaleships could be recognized by the wooden davits standing above the rails from which hung the active whaleboats. As a four-boat vessel, the *Clara Bell* had one pair of davits on the starboard (right) side near the stern and three pairs on the larboard (left) side, the first mate's aft, the second mate's amidships, and the third mate's forward. Each davit was fitted with a set of blocks and tackle for raising and lowering the boat. Below each davit was a triangular, hinged structure called a crane, which was swung out to support the weight of the whaleboat when it was hoisted into position on the davits and swung in when the boat was lowered.

Equipped with four whaleboats, the *Clara Bell* departed with a crew of 22 men, with the rest of the necessary hands to be recruited in the Azores. The initial crew included:

Charles H. Robbins, master
John C. Barker, first mate
David M. Welch, second mate
Franklin Perry, third mate
Manuel Joseph, fourth mate and boatsteerer
William C. Johnson, boatsteerer
Frank Joseph, boatsteerer
William Campbell, carpenter
John Hamcon, cooper
William Stevens, cook
James Lang, steward
Joseph Dutry, seaman
Manuel Pine, seaman
G. Francis Ladd, ordinary seaman
George B. Brigham, greenhand
C. E. Holt, greenhand
Mulford Howland, greenhand
John Joseph, greenhand
Henry Neagus, greenhand
Theodore S. Ransom, greenhand
Andre S. Rounds Jr., greenhand
Robert Wallace, greenhand

First Mate "John C. Barker" is an enigma. He may have been 28-year-old John E. Barker of Newport, Rhode Island, who had first gone whaling at age 14 on the bark *Ann Maria* of Fall River, Massachusetts, in 1843. Second Mate David M. Welch had gone to sea as a greenhand on board the *Montezuma* in 1846 and in 1853 had served as second mate of the *Tropic Bird*. Third Mate Franklin Perry had left his home on São Jorge, Azores, as a teenage greenhand and at age 20 had risen to boatsteerer on board the *Emma C. Jones* in 1848, then made that vessel's 1852 voyage as third mate. In a poem in text, Weir implies that the cook, or "doctor," William Stevens was Black.[11]

Weir brought what he called a "scratch journal" with him and, when duty permitted, he compiled his observations, his feelings, and descriptions of the whaling process in intermittent diary form. Soon his entries took the form of a logbook, following the sea day, which ran from noon to noon, noting wind and sea conditions and courses sailed, and including notations on whales sighted and barrels of oil taken from whales killed. Possibly on his return from sea, he stopped at nautical instrument maker John Kehew's shop on North Water Street in New Bedford and purchased a blank journal. Into this he would copy his daily remarks in more legible form and add illustrations in pen. Years later, he would copy the text of his "scratch journal" into the back in pencil.

What follows is a transcription of the journal, which, due to its illustrations and its breezy manner, is a favorite among the many logs and journals preserved in the G. W. Blunt White Library at Mystic Seaport Museum. The text has been lightly edited for ease in reading, with most abbreviations spelled out and misspellings corrected. In addition, contextual notes have been added after many entries to provide information on maritime and whaling practices mentioned, marine life encountered, places visited, and vessels mentioned.

Notes

1. The Weir family genealogy and Robert Walter Weir's artistic career are presented in most detail in Marian Wardle, ed., *The Weir Family, 1820–1920: Expanding the Traditions of American Art* (Provo, UT and Hanover, NH: Brigham Young University Museum of Art and University Press of New England, 2011).

2. Holy Innocents Episcopal Church, "History," https://holyinnocents2.wixsite.com/hihf/history, accessed November 2023. John F. Weir's comments on his father, in an 1864 letter to his future wife Mary, in the Weir papers at Yale University, is quoted in Wardle, ed., *The Weir Family*, 9–10.

3. Wardle, ed., *The Weir Family*, 9.

4. "History of West Point Foundry," https://web.archive.org/web/20070629101949/http://www.scenichudson.org/land_pres/wpfp_research.htm, accessed September 2023.

5. "Chadwick Family Papers," New York State Library, Albany, NY, https://www.nysl.nysed.gov/msscfa/sc16555.htm.

6. A period history of whaling, including a chronological listing of vessels, their masters, whaling grounds, and landings, is Alexander Starbuck, *History of the American Whale Fishery* (1876; reprint, Secaucus, NJ: Castle Books, 1989). A more recent source is Eric Jay Dolan, *Leviathan: The History of Whaling in America* (New York: W. W. Norton, 2007).

7. William B. Whitecar, *Four Years Aboard the Whaleship. Embracing Cruises in the Pacific, Atlantic, Indian, and Antarctic Oceans. In the Years 1855,'6,'7,'8,'9* (Philadelphia: J. B. Lippincott, 1860), 13–18.

8. *New Bedford Directory* (New Bedford, MA: Charles Taber & Company, 1856). Robert Wallace, Receipt, August 18, 1855, VFM 1743, Manuscripts Collection, G. W. Blunt White Library, Mystic Seaport Museum, Mystic, CT. The use of a number rather than a name to identify the month on Weir's receipt indicates that, like many Nantucket and New Bedford whaling merchants, James Barney and Charles Spooner were Quakers. Mystic Seaport Museum Curator Michael Dyer identifies the "Harris" to whom Weir was indebted for $14.00 as tailor Henry Harris, who produced seamen's clothing.

9. Charles H. Robbins, *The Gam: Being a Group of Whaling Stories* (1899; Salem, MA: Newcomb & Gauss, 1913).

10. Robbins, *The Gam*, 174.

11. Whaling Crew List Database, New Bedford Whaling Museum, https://www.whalingmuseum.org/online_exhibits/crewlist/search.php, accessed September–November 2023. Thanks to Mystic Seaport Museum Curator Michael Dyer for suggesting that John C. Barker was actually John E. Barker.

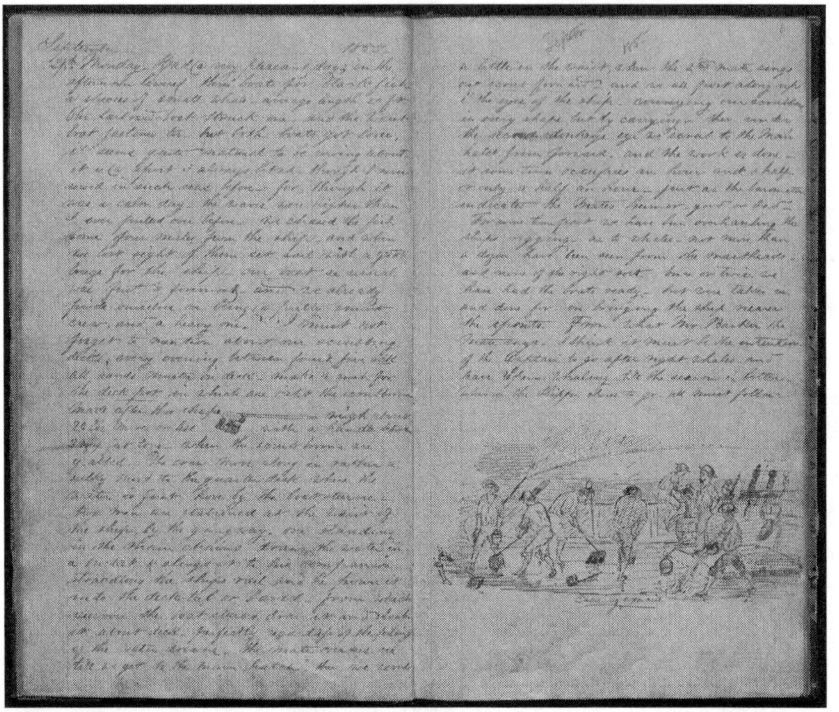

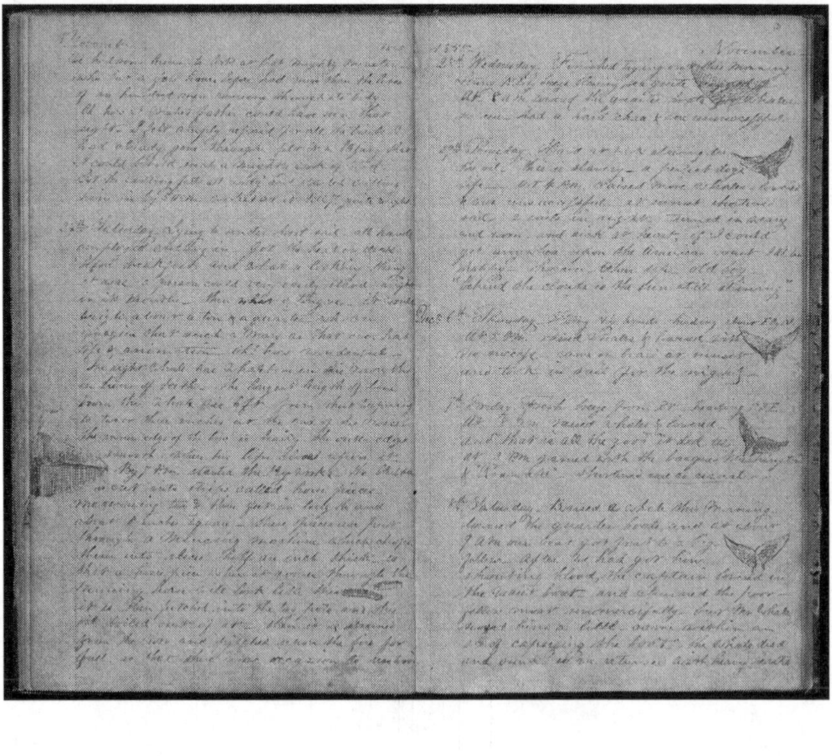

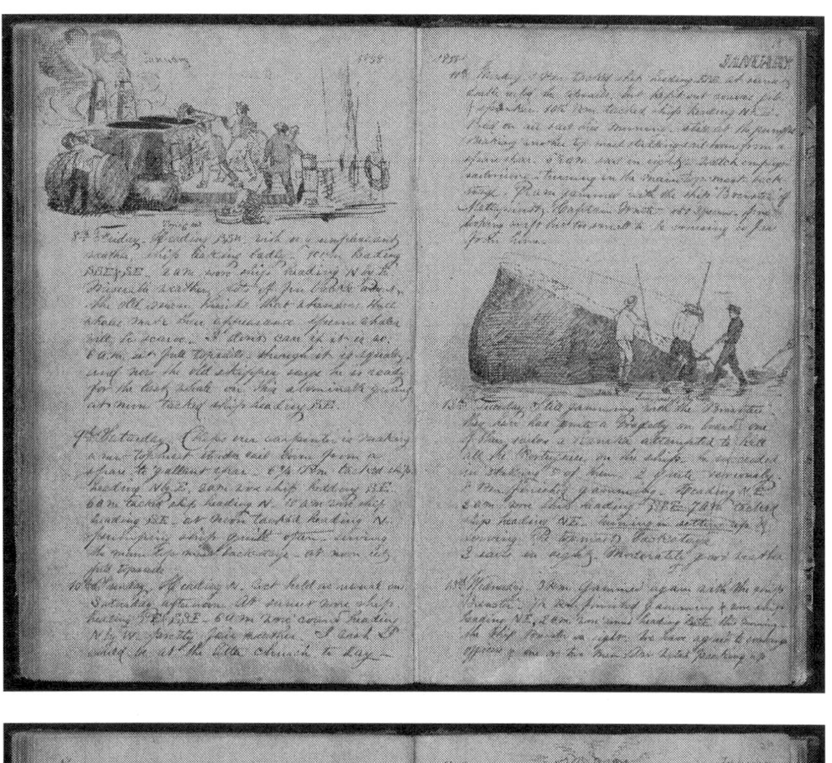

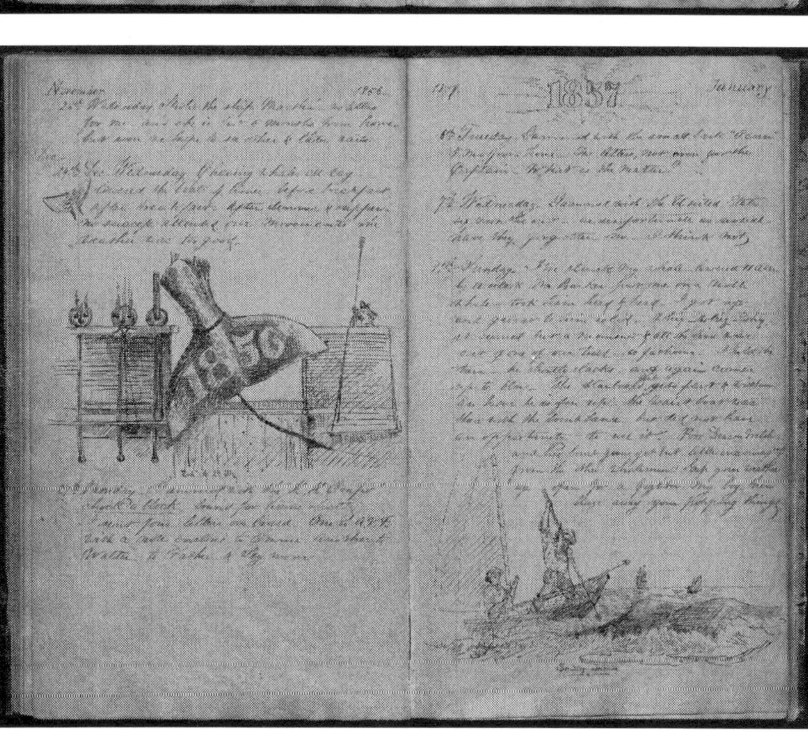

The Journal

1855

Saturday, August 18th, 1855

Hauled off from the most detestable of places, New Bedford, and here we are anchored about two miles down the stream. I am glad to get away from New Bedford. Never have I spent two such wretched weeks in all my life. True, I had a pleasant place to board at Mrs. Doan's, but I could never feel settled no matter how long I might stay there. I read every book and every piece of book that I could find. Tried to smoke plug tobacco pipes but couldn't enjoy that.

Sunday, August 19th

Oh! If the folks at home knew what a field I am about to launch upon, what would they say? What does dear father think? But I cannot turn back. I may just as well as not begin to cut my way in the world now—rather than leave it till I am older. I spent this day sacrilegiously in climbing about the rigging; didn't venture much but guess I'll soon get used to it. Hurrah for hard times. At least I'd like to make myself feel so, but I scarcely dare look ahead. It seems rather dark but I have great anticipations of future independence. I shall never, never call on father again, but I dare not speak his name. I have wronged him too much to be his son.

Having climbed as high as the fore-topgallant yard and studied the rigging, Weir noted in his scratch journal, "I am in for it now and no mistake."

Monday, August 20th

A day to be remembered. The captain came on board a little after 9 o'clock, and we weighed anchor and set sail. Then came the first touch of work, in handing up the anchor. Such a ponderous thing is only fit to be buried at the bottom of the sea. I sincerely hope we shall not have the pleasure of dropping it till we again reach home. The chains were soon stowed between decks, or rather in the chain pens, and the anchor catted and lashed. Now we are on our way rejoicing, and the first mate sent me aloft to slush the fore topgallant mast. In the afternoon the crew were divided into two watches, the larboard and starboard. I belong to the mate's or larboard watch.

Whaleships carried two anchors, a heavy "best bower" to starboard (right) and the similar-sized "small bower" to larboard (left). The heavy anchor chains ran through iron hawsepipes at the bow and around the windlass, which was turned by the crew to raise the anchors. When the anchors were at the surface, a heavy block and tackle was used to raise them to short, heavy timbers near the bow called catheads. Once an anchor was "catted," it was hoisted over the rail to lie on the small anchor deck at the bow. The chains were disconnected and run down to the chain locker or bin in the hold, where their weight helped serve as ballast. As Weir noted in his scratch journal, when ordered to slush the mast, "a sailor named Manuel, a Portuguese, kindly told me what to do." Grease derived from cooking salt pork and salt beef was used to lubricate the upper masts so the yards slid easily when they were raised or lowered as the sails were set or furled. To operate the vessel round the clock, the crew—excepting the "idlers," including the cooper, carpenter, cook, and steward—was divided into two work groups or "watches" of about eight men. The day was divided into four-hour watch periods, and the sea day began at noon. During daylight, everyone was involved in ship's work, but overnight, the watches alternated duty steering and handling sails as needed. So that one watch did not always have the midnight-to-four duty, the 4 to 8 PM period was divided into two "dog watches." During this period, each watch had two hours for supper and recreation. The first mate commanded the larboard watch while the second mate stood in for the captain to command the starboard watch.

Tuesday, August 21st

Beginning to get seasick and disgusted. Land is out of sight—I feel awful. We have to work like horses and live like pigs. My eyes are beginning to open to my rather dearly bought independence; however, we shall get on the sunny side shortly, I hope.

Wednesday, August 22nd

We are far, very far, out of sight of land, of sweet Ameriky. I was sent aloft on the lookout for whales and whatnots—and oh! How dreadfully sick I was! I saw two sharks, one about 12 feet long and the other 5 or 6 feet. I felt very much terrified to throw myself to them for food. I can truly say I never was disgusted before in my life. The sea presents a cobalt blue appearance—it is beautiful. In the afternoon, I took my first trick at the helm—two weary, dreary, desolate hours. Can a human being get toughened to all this?

Weir feigns an immigrant's pronunciation when using the term *Ameriky*. All vessels stationed a lookout at the bow to watch for danger ahead, but whaleships placed them high aloft to scan for spouts or other signs of whales. Each man in a watch would take a one- or two-hour stint at the helm or wheel, under the supervision of his mate. The mate would order him either to steer a compass course, watching the ship's compass mounted in front of the wheel, or to steer "full and by," watching the sails to keep them full while steering as close to the wind as possible.

Thursday, August 23rd

Sick as ever if not more so—but for all that, have to work like a dog. We have an excellent breeze but our bark pokes her nose under outrageously. I must not forget to mention we were all called aft by the captain before the pilot left us. At the same time the first and second mates picked their watches. Captain Robbins gave us a short harangue of which I noticed these few words: he'd give us plenty to eat and plenty to do, and if we acted like 'em, he'll treat us like men—no swearing, etc. etc. This afternoon the captain caught an albacore, a fish about 3 feet long, very thick and solid. I turn in, emphatically speaking, disgusted and thinking of home.

The pilot navigated the vessel out of New Bedford Harbor and through the islands and shoals until it was safely offshore, at which point he was picked up by a pilot schooner to be prepared to navigate another vessel back into New Bedford. It was a practical tradition for a whaling captain to address his crew at the beginning of a voyage to introduce himself, explain his expectations and the rules of the vessel, and perhaps to instill some common spirit, or fear of discipline, in the men. Captain Robbins later wrote a fictionalized account of an occurrence during this *Clara Bell* voyage and imagined the men describing him this way: "as thorough a seaman as ever trod the quarter-deck.

"Strict?

"Yes—an' no. You don't feel like you was being governed, an yit ev'ry man aboard done what he said, ev'ry time. They ain't no half-laughs an' sailor's grins about him. He's straight up an' down like a yard o' pump water.

"A jolly wag, too!" (Robbins, *The Gam*, 182.)

Captain Robbins caught an albacore tuna, possibly using the vessel's multi-pronged "grains," or fish spear.

Friday, August 24th

The day commenced with a very stiff breeze that increased so much we had to take in most of the sails. It rained pretty hard in the evening and I got wet and tired out tending the rigging and sails. I tumbled into my bunk with exhausted body and blistered hands. Romantic.

Saturday, August 25th

The wind still blows pretty hard and the decks are constantly washed by the waves. I have not quite recovered from sickness yet, but I think I am getting better. I am absolutely sick and disgusted with the living and everything.

Sunday, August 26th

Commenced the day at the masthead feeling quite well; while looking about for whales or rather nothing (for I did not search the seas much as it was the Sabbath) I had pleasant thought of those I left so unkindly and abruptly, but I console myself that it will be some relief to dear father for me to be off his hands. I also amused myself by singing all the psalms and hymns, chants etc. that dear Em and myself used to sing in our little church. By the time my patience was pretty well exhausted and seasickness beginning to come on, my relief came very leisurely up the rigging and now once more I find myself on deck, but I am so sick from the rocking of the mast that I cannot read much in my bible as I intended and can scarcely write.

About noon, it was a dead calm with heavy swell and the sails make a very loud noise flapping against the masts. I tumble into my bunk in disgust.

Other than ship-handling and keeping a lookout for whales, Sunday was normally a day of rest on a whaleship at sea. Men might "overhaul" or review the personal items in their sea chests, mend clothing, or, like Weir, read. Weir refers to his sister Emma.

Monday, August 27th

Good breeze blowing—another week of toil before I cheer up. I will soon get used to hard work and look at it as play, but the feed—awful. The waters have not been quiet enough to allow writing with ease since we started. Often a big lurch of the ship will knock half the ideas out of one's head. I must give up now anyhow.

My turn on lookout at the masthead from 1 o'clock until 2. While there I saw a school of cowfish and they appeared somewhat like the bodies of cows tumbling about in the water. I also saw plenty of flying fish—I never imagined there were half so many in the seas. I saw some land swallows, one or two of which lit in the rigging. They did not remain long—rested an hour or more and then went home happy creatures.

By 4 PM it was blowing quite a gale. Plenty of rain, wind still increasing, and both watches were sent aloft to take in sail. It must have been a rich sight to see us all scrambling up the shrouds. The ship was almost on her beam ends by the wind and the spray dashing nearly to the foretop several times. I have had to go aloft in rain and wind to furl and reef—and I do like it. There is not half the danger that landsmen imagine.

But at such times we can feel our utter insignificance. It was our midnight watch on deck, and by that time the rain had ceased, and the moon would occasionally peep from behind the clouds, making it look glorious. We were scudding before the wind, under double reefed topsails and at a pretty good speed. It seemed to me full 12 knots. I commenced there under the occasional light of the moon to appreciate something of the romance of the sea—the thought of which when I was seasick utterly disgusted me. I feel joy in beholding the Most Noble Ocean.

Whaleship food was very basic: boiled or pan-fried salt beef or salt pork, possibly boiled potatoes, hardtack (very hard crackers), possibly bean soup, occasionally sourdough bread, and coffee with molasses for sweetening. Vegetables, fruit, and fresh provisions were only available when a vessel called at an island or port. Weir uses the period term "cowfish" for the grampus, or Risso's dolphin, a marine mammal of the dolphin family that grows to about 13 feet in length. Represented by more than 40 species throughout the oceans, flying fish evolved long pectoral fins that act as wings for gliding. Possibly to evade predators, flying fish, which average about 10 inches in length, break the surface at up to 35 miles per hour and glide above the waves for up to 600 feet. When land birds like swallows were blown offshore, they might sight and land on a vessel to rest. The stays or shrouds supporting the mast sections from side to side were laced with ratlines to serve as ladders aloft. The crew would climb the windward side, which would become less steep as the vessel heeled away from the wind. "Beam ends" means heeled over so the deck is almost vertical. The topsails had three horizontal bands of short lines called reef points. To double-reef, the yard was lowered enough so the crew could go aloft and tie the middle band of reef points around the yard to reduce the area of the sail by about one half. A vessel's speed was measured in knots (see note for September 27). Twelve knots was quite fast for a vessel the length of the *Clara Bell*.

Tuesday, August 28th

Had a splendid breeze all day. At about 10 o'clock this morning, a poor, tired swaller lit in one of the davits. I presume it must have been blown from the far distant shore in the gale. I sympathized with the poor creature, fed it and before noon it flew away. We are now steering to a little south for the Azores, or as all on board call them, the Western Islands. This evening had quite a heavy blow which lasted till early this morning. At about 11 PM I saw a most beautiful rainbow—I never noticed one in my life before this.

Wednesday, August 29th

I did not trouble myself much with what transpired, for I was quite sick and exhausted and absolutely disgusted with life in the forecastle. But the beginning of fortune is later—keep up a stout heart, my boy—but wouldn't those at home pity me if they only knew my present situation? Ah well, they don't, and I intend they shall not.

This may have been the undated Wednesday when Weir wrote on the back of his outfitting receipt a tiny pencil inscription: "Oh what it is for me to leave that one for whom my heart is so anxiously & continuously longing." Presumably he refers to Anna Chadwick, though possibly to his sister Emma (VFM 1743, Manuscripts Collection, G. W. Blunt White Library, Mystic Seaport Museum).

Thursday, August 30th

For nine days we have been out of sight of land, and for the last four days, nothing has broken the line of the horizon—I haven't heard the cry,

"There she blows!" yet—but we are not left idle. Every day since we left home, the hold has been overhauled or something otherwise done about the ship—seizing on the chafing gear to prevent the rigging wearing away and other such innumerable jobs.

"A sailor's work is never done" runs the aphorism, and both to maintain the vessel and to prevent idleness, that was true. Casks in the hold were rearranged as food and water were removed, sails were checked and repaired, lines were checked and replaced, the standing rigging was seized (wrapped) with spun yarn and tarred, chaffing gear (sometimes called baggywrinkle) made up from short bits of old rope was attached to standing rigging where lines or sails might chafe against it, masts were slushed and woodwork was painted, and, sometimes, the hand-cranked rope machine was set up to turn fibers of old rope into spun yarn.

August 1855

It gives me great relaxation when I am on the lookout at the masthead to sing over the dear old songs that Emma and I used to amuse ourselves with at home—for father's amusement, visitors, and a great many too that could never have appreciated them.

Friday, August 31st

Yesterday we had a half hour's practice with the boats. I heard the call "Man the boats" while at the masthead, and down I had to scramble to be at my post for the larboard boat. All four boats were lowered, and after maneuvering about for practice within a mile of the vessel, going through all the motions of harpooning and avoiding the struck whale, we raced to the ship and our boat beats. The crew of each boat amounts to six men—the first, second or third mate, the harpooneer or boatsteerer, and four men. After the harpooners have fastened to a whale, he changes places with the mate, taking the steering oar, while the mate goes in the bow of the boat and uses the lance to kill the leviathan.

We never see such seas in our rivers, as there is the calmest day here—I may be mistaken but haven't found it different yet. Early this morning while standing my watch on deck, a flying fish came on board. The first I have had a chance to see one so closely—it was about nine inches long, and from tip to tip of its wings measured eight and a half inches.

So far, the crew have deported themselves very peacefully—nearly half of them are Portuguese.

I don't like them, though I can make out to live peacefully. Our mate is a villain—I can see it but too plain, but he must be careful. The second mate is a boy—the third mate a Gee.

Especially with so many greenhands on board, it was important to train them in all the intricacies of handing a whaleboat and develop the teamwork to do so under oar, sail, or paddle, backing at the order "stern all!", turning, and rowing as fast as possible. In a four-boat whaleship, the captain normally commanded the starboard boat, the first mate had charge of the larboard boat, the second mate commanded the waist boat, and the third mate directed the larboard boat. Each might have his own style of command and preferences for the operation of his boat. Weir expresses the common discriminatory views of other races or non-English-speaking peoples. "Gee" is a derogatory term for Portuguese, meaning to whalemen either Azoreans or Cabo Verdeans.

Saturday, September 1st

Sailing along with a pretty smart breeze towards evening, running along under double reef two fore and main topsails; by 7 PM commencement of my watch almost under bare poles, fore and main spencers, fore staysail and close reefed main topsail. The seas by this time rolled mountains high; occasionally our decks would be swept, but upon the whole our noble bark rode gallantly on. Oh God, how beautiful are all thy works—in wisdom hast thou made them all. What are we upon the bosom of this mighty ocean? Why can't we be more deserving of thy kind mercies, sailors and those who make the sea their home—can we not be Christians?

For truly we cannot help feeling how God continually watches us while on the stormy deep.

The topsail—the second square sail in ascending order on a mast—was the working sail, the first square sail set and the last furled. The *Clara Bell*'s topsails had three bands of reef points used to tie off—reef—either one-quarter (single reef), one-half (double reef), or about two-thirds (close reef) of the sail. The forestaysail was the innermost triangular jib at the bow, set on the forestay. Spencers were four-sided fore-and-aft sails set from a gaff on the after side of the fore- and mainmast. They were similar to the spanker on the mizzenmast but did not have a boom at the foot.

Sunday, September 2nd

Our Larboard Watch commenced again at 3 o'clock this morning—the sea was in glorious commotion. We would see an enormous wave come rolling toward us, with every prospect of being overwhelmed—but no, God is there—our vessel would glide gently over it, through a brilliant dash of spray and foam. By 9 PM the moon broke through the clouds and showed the scene in all its grandeur. Oh how wonderful art thou, Oh! Most merciful Father in all thy works, who can appreciate the beauties of Thy Lands!

By noon, the wind had subsided a little. I passed a most miserable Sabbath, for I could neither read nor write. I had to take my station aloft on the lookout, between 2 and 4 PM, though I did not go higher than the fore topgallant yard—the weather being bad! I was quite exhausted and faint from holding on, for the vessel tossed about most unmercifully, and very little did I see, like a whale.

Monday, September 3rd

Pretty heavy wind (and sea yet), though nothing in comparison to yesterday's grandeur—I think I'm beginning to get accustomed to a whaler's life, though if I get a chance to quit I'll do it. Tough times—might be easier.

Monday, September 17th

Two short weeks have elapsed since I last got a chance to jot any items—it has been all work and no play, and I have not till now felt able to do anything in the writing line with the exception of a bungling letter to Seymour—and 10 to 1 he never gets it. A great deal has happened during the fortnight above mentioned—we have laid off two ports in the Azores, Flores and Fayal.

At Flores we recruited ship—took on board any quantity of potatoes, pumpkins, onions, fowl, etc. Plenty of grapes—but to obtain them it is necessary to have plenty of money. With tobacco we made pretty good trades for apples, peaches, figs and cheeses—donkey cheeses at that—and right good and wholesome they were. I don't know that I ever enjoyed fruit and cheese so much—it seemed as though we had been deprived of them for years instead of a few weeks. It was long enough to make me long for something fresh, as I was so totally disgusted with ship fare. Am I now getting more used to salt junk, coffee, and tea? I can't say what the coffee is made of, but it resembles that delicious beverage as much as ink resembles water—there is but little coffee taste about it. Still, it is a little preferable to the stale water.

As for the tea, if I only had some of the currant leaves off the bushes in our front yard, I'd feel grateful—but there is no use crying over spilt milk—get case hardened and go ahead, but I would like someone at home to have a sip of this same tea or coffee.

Seymour is Louisa's husband, Truman Seymour. The nine volcanic islands of the Azores lie 2,000 miles east of New Bedford and 900 miles off the coast of Portugal. First colonized by the Portuguese in the 1430s, the Azores were also settled by refugees from Flanders in what is now Belgium. The islands became a calling point for New England whalemen in the 1700s and did a good business in supplying food. With population pressures throughout the islands, young Azorean men frequently joined whaleship crews and developed a proficiency for whaling. Flores was the largest island of the western group. Like the *Clara Bell*, whaleships commonly called at the principal port of Horta on the island of Faial, where the US consul lived.

Weir emphasizes the importance of fresh food to whalemen after their diet of salted meat.

Sunday, September 23rd

I am getting quite used to work now, and my hands can testify that quite plainly, for they are as hard as horn inside. Pulling and hauling on hard ropes—and the outsides have a most beautiful vandyke brown color—all in the voyage—the handling of ropes and tar has a very visible effect on the hands. I like the sea and I always thought I would—but it is hard to be separated from the dear ones at home by such a distance of waters.

The manner in which the Sabbath is spent on board by nearly all is truly deplorable; there is very little regard paid to the day if we are not making and taking in sail all the while. If the weather is pleasant, most of the men will be seen squatting about the fore hatch smoking, dozing or growling—some read, wash, saw, etc. etc.—but no thought is given to the welfare of the soul. And our noble first mate sets a most beautiful example by lounging on the quarterdeck, the picture of idleness and misery. If he could, he would like nothing better than to keep the crew hard at work Sundays as well as week days.

I shall indeed be thankful to get settled somewhere on land, where we can have a chance of improving the mind and choice of good companions. Nil desperandum.

Nil desperandum is Latin for "Do not despair."

Monday, September 24th

Had a very pleasant day; in the afternoon I lowered three boats for black fish—a species of small whale—average length 20 feet. Our larboard boat struck one—and the waist boat fastened too—but both boats got loose. It seems quite natural to be rowing about; it is a sport I always liked, though I never rowed in such seas before. For though it was a calm day, the waves were higher than I ever pulled over before. We chased the fish some four miles from the ship and when we lost sight of them, we set sail with a good breeze for the ship. Our boat as usual was first and foremost—and we already pride ourselves on being a pretty smart crew, and a heavy one.

I must not forget to mention about our scrubbing decks; every evening between four and five o'clock, all hands muster on deck and make a rush for the deck pot in which are kept the scrub brooms. They are made after this shape and weigh about 20 pounds more or less, with a handle between three and four feet long. When the scrub brooms are grabbed, the crew move along in rather a sulky mood to the quarterdeck where the water is first hove by the boat steerers. Two men are stationed at the waist of the ship by the gangway. One standing in the main chains draws the water in a bucket and slings it to his companion straddling the ship's rail, and he pours it into the deck tub or barrel—from which reservoir the boat steerers draw it and slosh it about deck, perfectly regardless of the feeling of the water drawers. The mate oversees us till we get to the main

"Evening Exercise" shows the daily ritual of scrubbing the deck as the sun goes down. Under supervision of First Mate Barker, standing on the main hatch cover at right, one man fills a barrel with seawater and others use buckets to spread it for those with heavy scrubbing brooms. Above them is the foot of the main course, with the tack line leading forward to the starboard rail.

hatch—then we scrub a little in the waist when the second mate sings out "Scrub forward"—and we all post up along the eyes of the ship, conveying our scrubbers in every shape but by carrying then under the second donkey's eye we scrub to the main hatch from forward—and the work is done. It sometimes occupies an hour and a half or only a half an hour—just as the barometer indicates the mate's humor—good or bad.

For some time past we have been overhauling the ship's rigging. As to whales, not more than a dozen have been seen from the mastheads—and none of the right sort. Once or twice we have had the boats ready, but were taken in and done for on bringing the ship nearer to the spouts. From what Mr. Barker, the mate, says, I think it must be the intention of the captain to go after right whales and leave sperm whaling till the season is better. Wherever the skipper choses to go—all must follow.

Blackfish was the whaleman's name for the pilot whale, a small-toothed whale that swims in pods. Blackfish may grow to 20 feet in length and were often hunted early in a whaling voyage for the boats' crews to practice. The scrubbing of decks was a daily routine. The main chains are the narrow timbers on the vessel's sides aligned with the mainmast that add to the width of the vessel to give the shrouds that are anchored there a better angle of support for the mainmast.

Thursday, September 27th

I have to take my chances for writing—a little less often than I could wish. We have not much breeze today and this morning by the Old Man's orders, hoisted the fore top mast studding sail. We have a fair breeze, but it is very light. All sail out and make less than three knots.

We are but little more than 20 days north of the equator—but we have had some pretty hot days already. Today it is quite calm and the sun pours upon us an overpowering mass of heat—for all that we must work, work, work, from sunrise till sunset.

Studdingsails were light sails with a small boom at head and foot, set outboard of the topsails and topgallant sails to expand the sail area for downwind sailing. A vessel's speed was calculated in knots, measured by a logline. Wound on a reel and having a triangular wooden "chip" on the end to anchor it on the water's surface, the logline would unreel as the vessel sailed on. Along the logline, knots were placed at either 42- or 48-foot intervals, and the line was allowed to run out for 14 or 28 seconds, measured by a sand glass. The number of knots that ran out indicated the number of nautical miles the vessel would sail in one hour. A nautical mile equals 1.15 land miles.

Monday, October 1st

Last Saturday spoke and gammed with the ship *Almira* clean like ourselves.

Sunday—land in sight and four sails. The land was San Antonio, one of the Cape De Verde Islands. We are now steering for the Tristan Islands.

The 362-ton bark *Almira* of Edgartown, Martha's Vineyard, commanded by Captain Crosby, departed for the Pacific whaling grounds on August 21, 1855, and returned on May 6, 1858, with 342 barrels of sperm whale oil, 2,000 barrels of right whale oil, and 3,700 pounds of whalebone (baleen). The ten Cabo Verde Islands lie 300 miles and more off the coast of Senegal, Africa. Settled by the Portuguese in the 1400s, they became a depot in the trade in enslaved Africans to Brazil and later as a supply depot for American whaleships. Young men from the islands, generally of African-Portuguese descent, also served on whaleships.

Tuesday, October 9th

Gammed with the ship *Golconda* of New Bedford. Captain Howland—three months out 120 bbls of sperm oil. We have lowered the boats several times for black fish (had previously lowered for drill) but were not successful.

The 331-ton ship *Golconda* of New Bedford, commanded by Captain Philip Howland, departed for the Indian Ocean whaling grounds on June 21, 1855, and returned in June 1859 with 1,467 barrels of sperm whale oil and 120 barrels of right whale oil. As largely solitary wanderers of remote seas, whaleship crews were eager to socialize when another whaleship appeared. Whalemen called this gamming. Normally, the captain of one vessel and his whaleboat crew would go aboard the other vessel, and that vessel's first mate and his whaleboat crew would board the first vessel. The officers gathered aft to share information and gossip, and the crews gathered forward, or in the foc's'le, to sing and swap tales, or "yarns." Captain Robbins later described a gam: "Two bells! One o'clock in the morning! And the ships were gamming still. The two captains were holding high converse in the *Clara Bell*'s cabin, and until the visiting master saw fit to return to his vessel, the visiting watch from the stranger made merry in the *Clara Bell*'s forecastle. It was a delight to see new faces. It was a rare treat to hear new voices. It was a fine, novel pleasure to match yarns all round.

"The men sat on the stout sea-chests along the sides of that semi-circular room in the whale-ship's bow. There were eighteen men in all; nine were hosts and nine were guests. Light streamed down from greasy lamps hung up on the bitts. The air was dim with the smoke of cheap tobacco. . . . On that particular occasion the *Clara Bell*'s forecastle had been a hilarious roistering place since seven in the evening. There had been songs and cards and smoke; and smoke and cards and songs. There had been long-spun, hair-lifting narratives of whaling adventures. There had also been news from home—some of it a year old, but still startling; and some of it six months new, every word an eye-opener" (Robbins, *The Gam*, 182–83).

Tuesday, October 18th

Lat 4°4' N, Long 19°21' W

Lowered the boats for black fish today and our boat was lucky enough to get one: we were fast for about half an hour and then hauled up and the mate killed him with the lance—but not till he had given us a sleigh ride, as the boys called it. And when he went into his flurry causing the water to foam and boil—that was cheering. This our first prize seemed to revive our drooping energies. Steering to the south and east.

In the original journal, "crossed the line"—meaning they crossed the equator—is penciled beside his drawing of the whale. Here, Weir begins to note the vessel's latitude and longitude, which he would have obtained from the mate or from the vessel's logbook. Often called a Nantucket sleigh ride, Weir refers to a whale swimming rapidly on the surface, towing the whaleboat. Sperm whales are capable of swimming at more than 20 miles per hour for up to an hour. The flurry was the whale's death throes, a lashing of the water before turning on its side, "fin out." Weir uses the convention of whaleship logbooks to record each whale taken with an illustration. Logbook keepers commonly used a wooden or whale-ivory stamp in the shape of the species taken, with the number of barrels of oil written in. Weir drew the whales and recorded the number of barrels on the drawing.

Saturday, October 20th

We have been enjoying pretty good weather of late, I have been kept about as busy as usual and feel as much disgusted as ever. One sail in sight today: hands employed in painting the boats. Those that are upon the cranes. We have seven boats; four on the cranes and ready for active service, and three over the house one of which is the Old Man's jolly boat. All captains at sea seem to enjoy this appellation—the Old Man.

That is out of hearing—but though the skipper may only be a master of a vessel—he considers himself insulted if not addressed as captain.

Whaleboats might be painted with distinctive bands of color so they could be identified. As Weir notes, in addition to the active boats, ready to lower, two spare whaleboats plus a smaller jolly boat for routine use were carried bottom-up on the skids, an elevated framework aft. Regardless of his age, a captain was called the "old man," a term sometimes of derision, sometimes of fond respect, by his crew in private.

Wednesday

We are now on the cruising ground not very far from Tristan D'Cunha, Inaccessible and Nightingale—where we expect to meet with right and sperm whales. This evening the watches were divided into three boats-crew watches and set. Prospects seem to brighten now that we are upon this celebrated whaling ground, but the weather is rather cool for the time of year.

Inaccessible and Nightingale Islands are part of the Tristan da Cunha group. When on the whaling grounds, to ensure that whaleboats could lower quickly, the ship's duty might be performed by individual boat's crews in successive watches rather than by the usual watches comprising half the crew.

Friday, November 9th

Pretty stiff breeze from the north and west. A little before 6 o'clock this morning I heard the cheering cry from mast head. "There blows!" "There blows!"—"There goes flukes." The captain was on deck in a moment and after singing out "where, away and how far off" jumped into the rigging with his glass—presently the lookout cries again "there blows" half a dozen times all is now excitement—a general rush is made to get a sight: when the Old Man sings out get the boats ready. Then there is a confusion, each boat's crew rushes to their respective boats and assist the boat steerers in preparing the boats for dropping. Two harpoons that are attached to the line one by a short line of four fathoms which is fastened by a loose bowline knot to the towline from the tub—and the other or first iron is made fast solid to the end of the tub line.

Presently the old skipper sings out haul aback the main yard and lower away—was done in a twinkling—the boats dropped into the water and manned as quick as the sail is set, oars shipped and off we go after the captain's directions. The whale is down now.

We are about half a mile from the ship—presently we see a flag floating from the mainmast truck—the whales are up—we see them from the boat—off we put—gain on them fast—get about three ships lengths when they lift their flukes and sound again—after having an exciting chase of three, four, or more hours we turn about and go on board disheartened—tired and disgusted—such is the sad history of the first whale we saw, chased and didn't get. This forenoon spoke and gammed with the ship *Olympia* of New Bedford eleven weeks out.

Weather was getting unpleasant—we took in sail at sunset as usual on whaling grounds.

As Weir describes, the whale line attached to the harpoon was coiled in a large tub in the boat. Those on board the *Clara Bell* could see the whales and used prearranged signals to communicate with the whaleboats, but still, whales often escaped. In his journal, Weir adopted the whaling logbook convention of using a whale's flukes to indicate a whale that escaped. In logbooks, a wooden or whale-ivory stamp was often used to mark the page. Weir drew a set of flukes each time, one of which was selected for use in this published version of the journal. The 296-ton ship *Olympia* of New Bedford, commanded by Captain John Ryan Jr., departed for the North Pacific whaling grounds on August 15, 1855, and returned on July 5, 1859, with 140 barrels of sperm whale oil, 1,321 barrels of right or bowhead oil, and 6,200 pounds of whalebone (baleen), having sent home 600 barrels of whale oil and 20,600 pounds of whalebone, probably from Hawaii.

Sunday, November 11th
Blowing quite a gale have been lying to for the past twelve hours—by evening wind died away a great deal heading east by south.

I wonder if the dear folk at home have any idea as to what part of the wide ocean I'm on now—shall I ever see them again.

Monday, November 12th
Raised whales a little after 7 AM, at 8 o'clock lowered the two quarter boats and shared like good fellows—but were unsuccessful. At 2 PM saw more whales and lowered again with as much success as formerly. At 4 PM gammed with the ship *Phoenix* of New Bedford—Captain Nicholsen 4 months out 36 bbls Sperm.

At sunset shortened sail.

The 432-ton ship *Phoenix* of New Bedford, commanded by Captain Horace Nickerson, departed on July 1, 1855, and would return on May 24, 1859. The vessel sent 255 barrels of sperm whale oil, 395 barrels of right or bowhead oil, and 19,400 pounds of whalebone (baleen) home, probably from Hawaii. The *Phoenix* returned empty and was sold to become a merchant vessel. When on the whaling grounds, a vessel would shorten sail at night to remain in the area.

Tuesday, November 13th
Blowing quite fresh from north and west at 7 AM lowered the larboard boat for a right whale; chased about half an hour and returned at 1 PM a whale again and the starboard boat lowered and chased without success—whale going fast to windward.

Thursday, November 15th
Inaccessible Island in sight—we went near enough to see the snow upon the summit of the peak—that makes me think of home sweet home, and that they will shortly be enveloped in the same kind of fleecy mantle.

Monday, November 19th
Gammed with the ship *India* at 1 PM. She is from New Bedford, Captain Howland 4 months 100 bbls sperm oil 175 whale oil—they have seen plenty of whales here and say it is a good ground.

The 366-ton ship *India* of New Bedford, commanded by Captain Timothy Howland, departed on July 18, 1855, and returned on June 17, 1858, with 963 barrels of sperm whale oil, 1,250 barrels of right or bowhead oil, and 6,200 pounds of whalebone (baleen), having sent home 35,000 pounds of baleen, probably from Hawaii.

Tuesday, November 20th
Today lowered for what was supposed to be sperm whales—chased about ten miles from the ship, quite calm. At 6 PM gammed with the *Almira* Captain Crosby—they said the whales we were chasing came near their vessel and were humpbacks.

Wednesday, November 21st
Lowered the larboard and waist boats for right whales—with our usual luck. If we keep on in this fashion we will be obliged to remain out three times three years to fill ship.

Thursday, November 22nd
Fresh breeze from the north and east with good weather. Lowered the boats three times this day for a right whale with our usual good luck. Beginning to get disheartened. Been having plenty of practice with handling the boats.

Friday, November 23rd
Hurrah! We have done it, we've taken our first whale and felt bad about it too—A little before 5 o'clock this morning just after the mastheads were set—the cry "there blows, blows, blows" soon got up the usual excitement—where away! Two points forward the lee beam—How far off! About two miles and a half—looks like a right whale—blows!

There blows! There goes flukes—get the boats ready my boys! We are now hard in chase with the ship the captain aloft with his glass—gives out his orders to the man at the wheel how to steer. The boats are now ready—we are nearing the whales—the courses are hauled up. The main yard aback—lower away—and the boats are in the water. At 5 o'clock the larboard and starboard boats were lowered and hard in chase—at about 5:30 the starboard boat got fast and away they dashed—passing within a ships length of our boat and going right straight to the windward—most of the time hidden in a cloud of spray. We immediately took in our sail and bent to our oars to get to their assistance but on and on they dashed madly jumping from wave to wave: the spray dashing high in the air: and often a heavy sea breaking over them for it was quite rough. It could not have been more than half an hour before they were out of sight from our boat, but we still kept pulling to windward in hopes the leviathan would mill round and come towards us and we might get a chance to put an iron into him. By 9 o'clock we had lost sight of the ship—she was then on the larboard tack. By about 10 o'clock she came in sight on the starboard tack lee rail under. Presently we saw that she had colors set at the three mast heads, signifying that she had lost sight of both boats. Again we saw them preparing to attack—they had hauled up the mainsail. When we thought it best to give them some signal to notify where we were, our sail or boats signal had been set some time—but they did not see that, so we loosed the sail and let it flap in the wind. They caught sight of it almost immediately and hoisted their signal at the mizzen peak for us to return on board and by 10:30 we had our boat safe on the cranes—but where was the starboard boat? All hands were sent aloft to look for it—and many were the disheartening conjectures that were made, soon we passed a broken oar—there is scarcely a breath heard from our mouths—a man's hat is passed, we are sick at heart—it is past 1 o'clock and no sign of the boat, our feelings by this time may well be imagined—at 2 PM Johnson the boatsteerer thought he saw something off our weather beam; the captain leveled the spy glass upon that speck—and immediately turned about and cried out get your dinner boys and breakfast too there's the boat alongside of the dead whale—what a shout of joy rang up from that ships company. We all felt an awful weight lifted from our hearts: before 3 PM we were

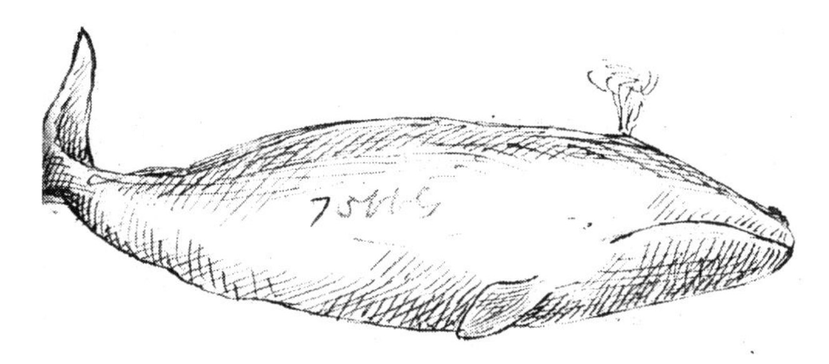

close to the boat and whale, they brought their line to us and we lay with head yards aback and hauled the whale alongside got the fluke-chain fast about his small and then we had some leisure to look at that mighty monster who but a few hours before had more than the lives of a hundred men coursing through its body. Oh, how I wished father could have seen that sight—I felt amply repaid for all the trials I had already gone through, felt it a blessing that I could behold such a mighty work of God. Got the cutting falls all ready and started cutting him in by 5 PM worked at it till 7 quite rough.

> Weir captures the frequent chaos of whaling. For "cutting in" to remove the whale's blubber and head matter, the carcass was fastened alongside, tail forward, with the fluke chain wrapped around the small of the body just forward of the flukes, passed through a hawsepipe, and secured on board. The cutting falls were sets of large blocks and tackle attached to the mainmast, with the hauling parts run forward to the windlass. The cutting falls could hoist the severed head, or parts of the head, on deck for processing. With a large iron blubber hook attached through the end of the strip of blubber called the blanket piece, the cutting falls were also used to bring the blubber aboard as it was cut free in a continuous spiral by the officers using cutting spades.

Saturday, November 24th

Lying to under short sail—all hands employed cutting in. Got the head on deck before breakfast and what a looking thing it was—a person could very easily stand upright in its mouth—then what a tongue—it would

weigh about a ton and a quarter, who can imagine that such a man as that ever had life and animation. Oh! How wonderful the right whale has whalebone in his mouth in lieu of teeth—the longest length of bone from this whale is six feet from that tapering to two or three inches at the end of his nose—the inner edge of the line is hairy—the outer edge smooth—where his lips close upon it. By 8 PM started the tryworks —the blubber is cut into strips called horse pieces measuring two and three feet in length and about eight inches square—these pieces are put through a mincing machine which chops them into slices half an inch thick—so that a horse piece when it comes through the mincing horse will look like this.

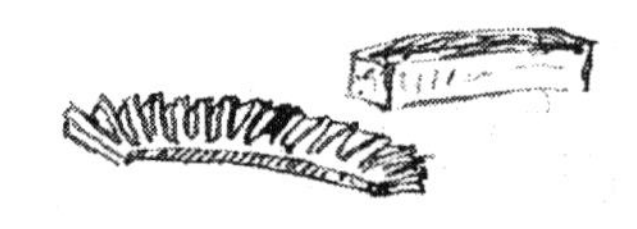

It is then pitched into the try pots and the oil boiled out of it—then it is strained from the pots and pitched upon the fire for fuel, so that there is no occasion to use wood.

The right whale (*Eubalaena*) was the original species hunted by American colonists. Growing to about 50 feet in length and perhaps 200,000 pounds in weight, about 40 percent of it blubber, right whales may live for 100 years. Females are mature at about nine years of age and breed every three to five years. Right whales are migratory but not fast-swimming, reaching only about six miles an hour. As filter feeders, they have about 500 plates or fronds of baleen hanging from the upper jaw, through which they force mouthfuls of water to strain out krill and copepods for food. With the dead whale alongside, the mates stood on small platforms beside the gangway and used long-handled cutting spades to remove the whale's upper jaw and then cut a spiral of blubber away from the carcass, which was hoisted aboard and cut into smaller pieces in the blubber room below before being minced for rendering in the tryworks.

Wednesday, November 28th
Finished trying out this morning. Strong northeast breeze blowing, sea quite rugged. At 8 AM we lowered the quarter boats for whales no use—had a hard chase and were unsuccessful.

Thursday, November 29th
Hard at work stowing down the oil—this is slavery—a perfect dog's life—at 4 PM raised more whales lowered and were unsuccessful. At sunset shortened sail—two sails in sight—turned in weary and worn and sick at heart, if I could get anywhere upon the American coast, I'll be happy—however—Cheer up—old boy—"behind the clouds is the sun still shining."

Thursday, December 6th

Strong north winds—heading about east by north. At 5 PM raised whales and lowered with no success—came on board at sunset and took in sail for the night.

Friday, December 7th

Fresh breeze from southwest heading southeast. At 8 AM raised whales and lowered and that is all the good it did us—at 3 PM gammed with the barks *Washington* and *Roanoke*—shortened sail as usual.

The 344-ton ship *Washington* of New Bedford, commanded by Captain Richard Holley, departed for the North Pacific whaling grounds on August 22, 1853, and returned on March 24, 1857, with 55 barrels of sperm whale oil, 1,802 barrels of right whale oil, and 15,200 pounds of whalebone (baleen), having sent home 18 barrels of sperm whale oil, 880 barrels of right whale oil, and 21,833 pounds of baleen. The 252-ton bark *Roanoke* of Greenport, New York, commanded by Captain Wade, departed on October 10, 1854, and returned on March 18, 1857, with 488 barrels of sperm whale oil and 351 barrels of right whale oil.

Saturday, December 8th

Raised a whale this morning lowered the quarter boats, and at about 9 AM our boat got fast to a big fellow. After we had got him spouting blood, the captain lowered in the waist boat and churned the poor fellow most unmercifully—but Mr. Whale nosed him a little—came within an ace of capsizing the boat—the whale died and sunk so we returned with heavy hearts.

The quarter boats were the starboard and the larboard boats, which were suspended from davits along the portion of the vessel aft of the mainmast rigging, which was called the quarters. The waist boat was carried on the larboard side along the waist, or midship part of the hull. The captain normally commanded the starboard boat. "Churning" means aggressively plunging and twisting the sharp head of the lance around the heart and lungs to mortally wound the whale. Both right whales and sperm whales usually floated after being killed, partly buoyed by their blubber. However, sometimes they did not have buoyancy and sank, which was commonly the case with humpback whales and other rorquals.

Sunday, December 9th

Lat 36°51' S, Long 10°53' W

At 6 AM we raised whales and lowered without success. At 9 AM lowered the quarter boats again for a right whale and Mr. Perry got fast to a noble fellow—we had not much difficulty in working around him but

The quarter boats approach a right whale, with its vertical spout. The boat at right rides up on the whale as the boatsteerer darts his iron and the crew peak their oars to prepare to manage the whale line.

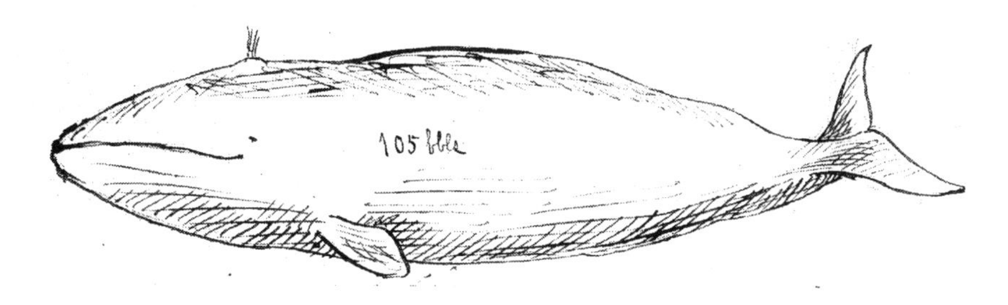

he did throw his flukes about most unmercifully—by noon he was fin up—and by 7:30 PM that monstrous leviathan could only be remembered by the pieces of blubber about deck and in the blubber room. This was a big whale; the thinnest part of a blanket piece measured about eight inches—some of the blubber was two feet deep. The right whale is a very dirty mammal compared to others of the same tribe—I have noticed they are covered with small insects very much resembling crabs about half an inch in diameter—on the end of their nose is a bunch of barnacles about 18 inches wide—this the whalemen call his bonnet—and when it has the appearance of a rock, the barnacles are enormous—as much as two inches deep—the boys often roast them and eat them the same as oysters—and many other tit bits do they have when a whale is "trying out" cooking—whale lean etc. etc.

Right whales have raised, callous-like patches on their heads, and whale lice—shrimp-like crustaceans called *Cyamidae* that look like crabs—cling among them to consume algae, skin flakes, and damaged flesh. A single whale may carry 7,000 whale lice. Whale barnacles commonly attach to the heads of right whales as well. A form of crown-shaped acorn barnacle, the whale barnacle imbeds itself and filter-feeds from the passing water as the whale swims, living for about a year.

Tuesday, December 11th

Hard at work trying out, and such a filthy looking set of men could not be met with on shore. At noon raised whales, and by 2 PM lowered and chased to no purpose.

Saturday, December 15th

Stowed the oil, 105 bbls last Thursday—an awful job had to break out clean to the keelson—chocked empty casks and ran the oil through a hose into them. This makes way with a great deal of labor—At 4:30 PM lowered for a whale and chased till sunset with usual luck. There are quite a number of whales on this ground, and we are most always in sight of one another four sails in sight at sunset.

Sunday, December 16th

Strong wind from the northwest running to the southward. Raised whales and at 3:30 lowered the quarter boats. Mr. Perry got fast—we pulled to the rescue and in a few minutes, Johnson had got his two harpoons solid and fast into the whale's side in less than an hour he was fin up.

Monday, December 17th

At noon lowered the quarter boats for whales. It was quite rough, and our boat got right in the center of a dozen barnacled noses—we were paddling, and therefore facing the whales—one whale came up directly under the boat—there was no alternative—he would lift us any way so soon as his flukes raised—so Manuel the boatsteerer pushed one iron into his back when quicker 'n Jarvis the boat was lifted almost clear of the water.

The bow's at an angle of 45 degrees all hands were thrown into the stern sheets and paddles and oars flew about in every direction—the whale darted off like mad—it was impossible to haul in any line for at least an hour—and the other boat's crew had a hard tug and after all could not succeed in fastening—so the captain called them on board, and kept the ship beating after us. This fellow blew with a terrific noise it sounded

like the bellowing of a thousand bulls—or like the exhaust steam from an engine. We now begun to haul line and even discovered a companion to our whale—we hauled close and found it impossible to get a lance, for the monsters dealt such heavy blows with their flukes that it would be sure destruction to go within a dart—again we hauled up and Mr. Barker gave the loose whale a lance which sent him puffing like a huge propeller close to the ship—the weather had now begun to be quite stormy and finding we could not haul up close enough for a lance, Mr. Barker took the boat spade and darted with good effect into the whale's flukes, it slackened his speed a little, though not enough to haul close—his flukes got pretty well cut up so that he left a river of blood behind him. It was now about 3 o'clock and the ship was almost out of sight to leeward and right close by us was the bark *Helen Augusta* but Mr. Whale took it into his head about his time to mill and run off for the *Clara*—he brought us within a quarter of a mile of the ship when the captain seeing but a poor chance for that whale put up the signals for us to cut and come on board—we were loath to do so though it was blowing a gale by this time. By 4 PM we were safe on board drenched to the skin, cold and shivering and had to take in sail and lay to: the decks were in an awful condition. While our boat was fast, the remainder of the crew had finished cutting in—the decks were slippery enough strewed here and there with pieces of blubber and the tongue was on deck sloshing about here and there—it was some time before it could be secured—then the blubber room casks were lashed in villainous confusion. When we had finished taking in sail, the captain sent all hands below to get some rest while he stood watch alone till about 6 o'clock when we had to come on deck and wear ship to avoid the land. Blew quite a steady gale all night—felt home sick. I wonder if I shall ever see my dear father again. My sisters—if they could see me now, how would they feel? Never mind cheer up! The time may yet come when I shall laugh at all these scrapes.

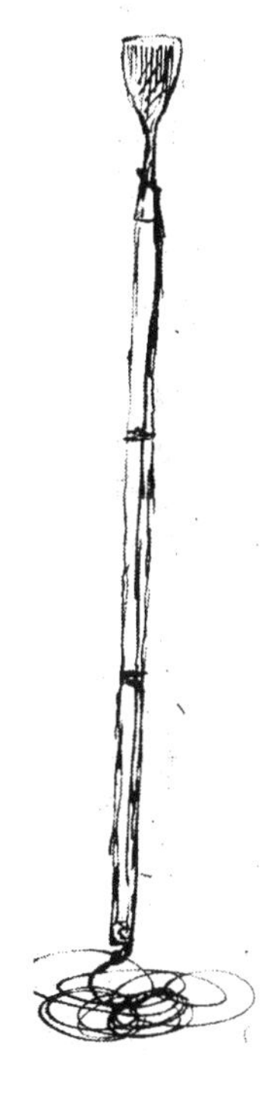

"Manuel" is fourth mate and boatsteerer Manuel Joseph. "Quicker than Jarvis" means immediately, but the origin of the phrase is not clear. It appears that another whale tried to protect the harpooned one. The boat spade was a short-handled version of the sharp-edged cutting spades used for cutting in. In the boat, it was used to cut a hole to attach a towline to a dead whale, but it could be used as Mate Barker did, to try to cripple the whale by cutting its flukes or tendons. The 270-ton bark *Helen Augusta* of Holmes' Hole, Massachusetts, commanded by Captain West, departed for the Atlantic whaling grounds on May 24, 1854, and returned on March 25, 1856, with 215 barrels of sperm whale oil, 890 barrels of right whale oil, and 2,000 pounds of whalebone (baleen), having sent home 3,000 pounds of baleen.

Thursday, December 20th

Strong northeast gale blowing. Lying under short sail trying out—rather disgusting work in such rough weather—bad enough in a calm. This morning raised whales—at 7 o'clock lowered the starboard boat and chased without a particle of success. Such weather as this is there seems to be more danger in leaving and returning to the ship than in being fast to a whale.

Friday, December 28th and Saturday, December 29th

Laying off on the Island of "Tristan D'Cunha" took on board about 40 bushels of potatoes with a few cabbages, turnips, carrots etc. I enjoyed quite a run on shore here—it is a very wild, romantic looking island—about 60 inhabitants—sent a letter to Emma by *Twilight*.

About 5,500 miles south of the Azores and 4,000 miles south of Cabo Verde, Tristan da Cunha in the South Atlantic was sighted by Portuguese mariners in the early 1500s but not settled until 1810. Tristan and its outer islands were annexed by Great Britain and became another supply base for American whaleships. With no harbor, vessels remained under sail and sent boats ashore. The 386-ton ship *Twilight* of New Bedford, commanded by Captain Isaac B. Tompkins, departed for the Indian Ocean whaling grounds on July 20, 1854, and returned on April 6, 1858, with 1,330 barrels of sperm whale oil, 127 barrels of right whale oil, and 1,000 pounds of whalebone (baleen).

Sunday, December 30th

Quite a calm day, about 18 sails in sight some have their boats down after whales. After dinner raised a right whale chased by killer whales—seemed to be sorely puzzled what to do with himself. Our boat was lowered with a picked crew and the second mate went as harpooner—we pulled onto the whale immediately and he lay almost double around the boat—Mr. Welch hove two irons—but pitch poled them—hauled in and we got the boat right over his small and he darted again with the same effect—the man was more frightened than the whale—and there were a dozen ships witnesses of our skill. I felt cheap enough though but a foremost hand.

Christmas day was not noticed on board—it happened to be my midnight trick at the wheel Christmas eve and if I did not feel bad, then I never did and never will—hard life this, but may get used to it—I forgot to jot down all the items of interest. There are numbers of seabirds that accompany us while cutting or killing whales among which the Albatross or Gooney is the most abundant—when there is a whale alongside, we find no difficulty in catching them. The largest we got measured twelve feet from tip to tip—and I believe they seldom exceed that size. Their feathers are beautiful, and they have a great quantity of them. It is wonderful to see this bird fly—it sails along so majestically, not making the slightest perceptible motion of its wings soaring high or skimming the waters—you seldom see them flap their wings except when rising from the sea.

The killer whale, or orca, the largest member of the dolphin family, is the principal predator for young sperm whales and other whales. Orcas live and hunt in pods. Second Mate David M. Welch seems to have gone to sea as a greenhand on board the bark *Montezuma* in 1846. By 1853 he was second mate of the bark *Tropic Bird*, and he remained in that role in the *Clara Bell*. Welch seems to have had a preference for the slightly safer bomb lance gun rather than the long hand lance for killing whales, and he was an embarrassing failure as harpooner in this case, causing the harpoons to flip over, or pitchpole, when he threw them, striking with the butt rather than the head. Several species of albatross are common across the higher southern latitudes and in the North Pacific. Among the largest of birds, the albatross wingspan may reach 11 feet. The birds nest on land but fly and glide for long distances at sea to consume fish, squid, crustaceans, and dead flesh. They use eyesight and smell to locate food and were probably drawn to a whaleship cutting in by the smell of flesh. Gooney was a sailor's term for the albatross.

1856

Tuesday, January 1st

What gay times they will have at home today—I wonder if they'll think of me. I must console myself by imagining they will. Last Tuesday Christmas day was scarcely noticed, in the afternoon we gammed with the bark *Helen Augusta* of Tisbury Captain West, 19 months out, 1,050 bbls right and sperm. Sent a letter home by her.

Wednesday, January 2nd

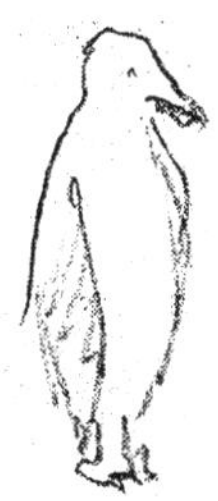

Strong breeze from the north and east heading east on larboard tack. Raised whales this morning and lowered the quarter boats by 6 o'clock, returned with our usual good luck. Caught a "Waugin" a species of sea bird that very much resembles the penguins—this was about the size of a drake. We are now bound for a cruise off the Congo River and St. Helena ground for sperm whales.

Waugin or woggin seems to have been a whaleman's term for one or more species of penguin in the South Atlantic and Indian Oceans and in the Northern Hemisphere perhaps for the Great Auk. (Storrs L. Olson and Judith N. Lund, "Whalers and woggins: A new vocabulary for interpreting some early accounts of the great auk and penguins," *Archives of Natural History* 34, 1 [2007]: 69–78.) Inshore along the African coast, from Angola up beyond the mouth of the Congo River to the Bight of Benin, right and sperm whales had been hunted in season from the late 1700s.

Saturday, February 23rd
Gammed with the bark *Mattapoisett* of West Port—4 months out 90 bbls sperm. Since leaving the Tristan ground we have had very good weather—hope soon to fall in with sperm whales off Walwich Bay.

The 150-ton bark *Mattapoisett* of Westport, Massachusetts, commanded by Captain James M. Sowle, departed for the Atlantic whaling grounds on October 23, 1855, and returned on April 11, 1857, with 570 barrels of sperm whale oil and four barrels of right whale oil. Weir refers to Walvis Bay on the coast of Namibia.

Tuesday, February 26th
At 4 PM gammed with the bark *Mary Ann* Fairhaven—33 months out 80 bbls Sperm. As she is bound home, we sent our bone on board about 1800 pounds. By 9 PM finished gamming—Captain Mccumber of this craft had his wife with him; a disgusting woman.

The 335-ton ship *Mary Ann* of Fairhaven, Massachusetts, departed for the Pacific whaling grounds on September 16, 1854, and returned on April 1, 1858, with 1,520 barrels of sperm whale oil. As the length of whaling voyages increased, captains began to obtain the owners' permission to take their wives and even children with them to sea. Some owners objected to the presence of a wife as a distraction to the business of whaling, but others found a benefit in it. "There is more decency on board when there is a woman," concluded whaling agent Charles W. Morgan in 1849. When Captain Robbins took command of the bark *Thomas Pope* after this voyage, he would bring his wife with him, and their daughter would be born on board. Overall, more than 400 captains' wives made more than 600 whaling voyages. (See Joan Druett, ed., *"She Was a Sister Sailor": The Whaling Journal of Mary Brewster, 1845–1851* (Mystic, CT: Mystic Seaport Museum, 1992) for a list of whaling wives at sea.)

Thursday, February 28th
Gammed all day with the brig *March* 200 bbls sperm, haven't spent such a lazy day since we sailed.

The small, 89-ton brig *March* of Mattapoisett, Massachusetts, commanded by Captain Henry Lewis, departed for the Atlantic whaling grounds on June 21, 1855, and returned on August 12, 1856, with 246 barrels of sperm whale oil and 29 barrels of right whale oil, having sent home 35 barrels of sperm whale oil.

Saturday, March 1st
Been racing all day with the bark *Sacramento* and gammed this evening—she is 16 months out 500 bbls sperm. She is a match for us—though if the breeze had been a little stiffer she could scarcely have kept up so well. The rumor afloat is that we are bound direct for St. Helena where we shall send home our oil and then go round the Cape of Good Hope to the Indian Ocean. It seems we are far enough out of the world now—and for my part I wish I was home again—but we may yet make a good voyage—cheer up.

The 218-ton bark *Sacramento* of Westport, Massachusetts, commanded by Captain Otis Snow, departed for the Indian and Pacific Ocean whaling grounds on November 10, 1854, and returned on April 16, 1857, with 337 barrels of sperm whale oil and 69 barrels of right whale oil, having sent home 125 barrels of sperm whale oil.

Wednesday, March 12th
At daylight this morning raised St. Helena a little on our lee—dropped the mud hook a little before 10 o'clock among about 20 sails merchantmen and whalers—had to turn to right off and work till sunset—breaking out shooks. It seems strange to have the sails furled again and have the vessel lie so still upon the water. We are anchored right opposite St.

Jamestown. The Island looks quite barren from the sea, but in the interior it is said to be quite fertile. I hope shortly to see for myself.

Discovered by Portuguese mariners early in the 1500s, the volcanic island of St. Helena lies near the middle of the South Atlantic Ocean, 1,200 miles west of South Africa and 2,500 miles east of South America. The island was well wooded and had abundant fresh water, so the Portuguese introduced livestock and crops. In the 1650s, the British East India Company turned the island into a supply base for the company's vessels sailing between England and India and China. In 1833 the island became a crown colony. By the mid-1800s, a thousand ships called at the island each year to obtain food and fresh water. Jamestown is on the island's northwest coast.

Tuesday, March 25th

Weighed anchor between 9 and 10 AM. Strong southeast breeze blowing with occasional squalls. I have sent two letters home from here—one to Walter in the bark *Afton* and one to Emma in the ship *Lancer*—We shipped all our right whale oil on the bark *Afton* she is bound directly home. From the time we dropped anchor till now we have had hard work—broke out all the oil 250 bbls for the *Afton*—stowed and filled 200 bbls of fresh water. The water boat came alongside and pumped water through a hose into our hold, painted the ship outside etc. I do not expect to have any more chances to send letters home, as we are bound direct for the Indian Ocean to cruise off Fort Dauphin Madagascar.

Weir refers to his older brother Walter and his favorite sister, now Emma Weir Casey. The 249-ton bark *Afton* of New Bedford, commanded by Captain James M. Clark, departed for the Atlantic and Indian Ocean whaling grounds on May 26, 1856, and returned on August 28, 1858, with 756 barrels of sperm whale oil, having sent home 120 barrels of sperm whale oil. The 395-ton ship *Lancer* of New Bedford, commanded by Captain Aaron A. Cushman, departed for the Pacific whaling grounds on August 4, 1856, and returned on

June 30, 1860, with 1,539 barrels of sperm whale oil and seven barrels of right whale oil, having sent home 54 barrels of sperm whale oil. Captain Cushman would die at sea on November 23, 1856.

Thursday, March 27th

Strong head winds oppose us. We are now on the larboard tack heading southwest by south or so. While at St. Helena, each watch had four days liberty—three days of which we each received one dollar from the captain—upon which large sum we could spree out 24 hours—and not get any bricks in our hats. There was considerable grumbling about the captain's generosity! Though we must certainly know it is for our benefit to draw lightly during the voyage for a better pocket full on returning home.

I visited Napoleon's tomb—it is about eight feet by five by ten deep, merely a walled hole in the earth with a dozen steps to descend to the bottom. It is in a beautiful spot, and nearby is a delicious spring by which I wish I could at this moment sit down. I have come to the conclusion that the land is the best after all—for at sea you never can be quiet—and must put up with all sorts of characters. Give me a home on the solid land, with a fairy to love me, and other dear ones that care for me—that would be happiness. Here I am often left to my own thoughts and they are not always very pleasant.

It is quite often that I have a longing desire to be again in my own native land—for I am inclined to think there are some there who care something for me—though I don't deserve much love certainly. Ah! The die is cast—and for two weary years and more I must be knocked about at the mercy of wind and wave before I can think of going home. Then

"St. Helena from Our Anchorage" shows vessel anchored off Jamestown. At right is Fort Hill, with the steps and rails leading up the steep slope.

when our vessel does go to that dear land who can tell what happy mortals we shall be but I am looking too far ahead—we know nothing beyond this minute. God only knows what the future may unfold. He is merciful. Before I close for tonight I must finish a few more remarks on St. Helena that far famed Isle. There were about 20 sail at anchor while we were there—nearly half of whom were American whalers; among the remainder were a beautiful British Iron Clipper ship as perfect a model as one would wish to see—she was painted white from stem to stern, with a beautiful white eagle for a figurehead, carrying a fore, main and mizzen sky sail. Among the whalers was the ship *Lancer* first from Fort Dauphin and filled "chock a block"—*Aerial, Plover, George, Dunbarton, Barclay* and *Marcela,* the *Afton* and one or two others left two or three days before us. There are several kinds of fruit upon the island but it does not seem to be very abundant. Pears and peaches predominate, but bananas, plantains, figs, grapes, apples, banyons etc. etc. are occasionally seen in the market. The day before we sailed, the Old Man sent on board 50 heads of cabbages, some turnips and carrots, 200 pounds of beef with sundry dainties for the benefit of himself and officers. Since we left Port, the weather has been bad.

Weir took $2.00 on advance, which in his final tally he called "wasted at St. Helena." In 1815, after the Battle of Waterloo and Napoleon Bonaparte's second abdication as emperor of France, the British government took the island as the most remote place to hold Napoleon in exile (as he had escaped from the Mediterranean island of Elba in 1814). A mansion called Longwood House was constructed, and Napoleon was taken to St. Helena in October 1815. He would live there until his death in May 1821. In 1840, Napoleon's body was returned to France and entombed in Paris. The iron clipper ship was the 770-ton, 210-foot *Lord of the Isles,* built at Greenock, Scotland, in 1853. It was the only iron clipper of the 1850s. *The Lord of the Isles* was a tea clipper, and in 1856 it won the race from the coast of China to London to deliver the first cargo of tea that year, making the passage in 127 days. The 225-ton bark *Aerial* of Fall River,

Massachusetts, commanded by Captain Borden, departed for the Indian Ocean whaling grounds on December 13, 1853, and returned on November 4, 1856, with 30 barrels of right whale oil, having sent home 326 barrels of sperm whale oil. The 330-ton ship *Plover* of New Bedford, commanded by Captain Charles M. Skiff, departed for the Pacific whaling grounds on August 28, 1855, and returned on February 27, 1857, with 171 barrels of sperm whale oil. The 280-ton ship *George* of New Bedford, commanded by Captain Jonathan Jenney, departed for the Atlantic and Indian Ocean whaling grounds on September 20, 1853, and returned on August 2, 1857, with 42 barrels of sperm whale oil and 937 barrels of right whale oil, having sent home 54 barrels of sperm whale oil and 9,000 pounds of baleen. The 199-ton bark *Dunbarton* of New Bedford, commanded by Captain Joseph P. Nye, departed for the Atlantic whaling grounds on November 14, 1855, and returned on August 22, 1858, with 237 barrels of sperm whale oil and 20 barrels of right whale oil, having sent home 160 barrels of sperm whale oil. The 281-ton ship *Barclay* of New Bedford, commanded by Captain Andrew J. Fuller, departed for the Atlantic and Indian Ocean whaling grounds, on August 1, 1854, and returned on August 24, 1857, with 410 barrels of sperm whale oil, 1,016 barrels of right whale oil, and 2,100 pounds of baleen. The 210-ton bark *Marcella* of New Bedford, commanded by Captain Benjamin S. Morton, departed for the Pacific whaling grounds on November 23, 1853, and returned on July 11, 1856, with 234 barrels of sperm whale oil, having sent home 63 barrels of sperm whale oil.

Sunday, March 30th

Five days from St. Helena had very stormy weather all the while steering full and by—heading anywhere between southwest and south. I am again getting used to sea life and do not care to go on shore again till we reach our own dear Native land—so I think now, but may alter my opinion. I dare not anticipate too much for the time is too far off, I wish we could fill up ship and start for home in the shortest possible time. I had a strange dream of home last night and of Planck's funeral taking placc, at the same time great festivities going on and Mrs. Cuyler being there. It is sadly unpleasant for me to have such dreams of home—though it is the second of the kind I have dreamed since leaving. It is seldom I dream unless unwell as last night, for all the sleep we can get it is too much needed to be frittered away in dreams. Oh! How I wish I could hear from home, what a weight would be lifted from my mind, what joy it would be to see all the dear ones again just as I left them seven months ago—is such a meeting in store for me? My dear father, does he think of me as his now? Oh! How deeply I have wronged you my father—can I ever be forgiven?—Oh how I wish I had always spoken with the freedom of a son to you—I might not have been here—but under such circumstances this may all be for the best.

While in St. Helena I attended church—and oh what happiness it was to be again in the house of God—after so long a separation from religious service—those moments I think, were the happiest I have had since we sailed. Our dear little church is often before me in my day dreams of home and the Hudson. Many a wondering thought do I give as to who is there and who not, but my thoughts are not enough to satisfy the craving desire to know something of those at West Point. I anxiously await more substantial intelligence and hope but a few months will bring relief.

I forgot to mention that while in port I ascended "Ladder Hill" by the steps—648 in number. It was quite a fatiguing task but I accomplished it, notwithstanding I had been rambling among the hills all the morning. These steps are built of wood upon a stone paved path, here and there are scattered wooden rollers, similar to those formerly on the road to old Fort Put. And at the head of the hill is a huge windlass which was used for heaving up the cannon, worked by mules.

Mr. Planck and Mrs. Cuyler were residents of West Point or members of the Weirs' church. Fort Put is Fort Putnam, a Revolutionary War fort at West Point that was in ruins during Weir's childhood. His father painted a scene of the fort and the Hudson beyond, which was published in 1831 as an engraving in William Cullen Bryant's *The American Landscape*. Ladder Hill rises 600 feet immediately behind the narrow settlement of Jamestown. To supply Ladder Hill Fort on the summit, the "Jacob's Ladder" stairway between a set of inclined plane railway rails was built in the 1820s.

Tuesday, April 1st

Rather pleasant this morning, but this afternoon was remarkably unpleasant. The wind died away at about 11 AM and since dinner it has been raining, not very heavy nor very light—a steady rain that will soak every rag one has on in less than five minutes. It is now 4:30 PM and it has cleared off finely with a stiff breeze from the north.

I did not scrub decks this afternoon—wonderful occurrence, perhaps the late rain saved us the labor, however I am obliged to jot this down, for either in rain or shine we have always had to scrub—unless it might have been very rough.

Wednesday, April 2nd

We are now in Latitude 24 south, and during the nights the weather is quite cool, so that it is really quite unpleasant standing watches. We may expect a rather poor time passing the Cape for none of us has very appropriate clothing for cool weather. All my spare time in a watch below is employed in sewing patches upon my clothes—for there is no disgrace in wearing patches at sea. I wish my sisters could see me in my whaling rig, they'd laugh some I'll wager. In the course of a few months, I presume my garments will be a mass of rags and patches, and I don't know, but it would pay to keep one pair of pants to present to Carnum or some such notorious humbug. No doubt they would be as great a curiosity as I shall obtain during the whole voyage—time will show.

Weir refers to the Cape of Good Hope, at the southern tip of Africa.

Thursday, April 3rd

Were it on land I should call this a beautiful day—light breeze and the temperature of the atmosphere is delightful—but such days as these we can't enjoy upon the sea, anything below a five-knot breeze is dull. This morning we again bent the main top-mast and topgallant staysails on temporary stays—preparatory to rounding the Cape. These sails were unbent the morning that we came in sight of St. Helena—why, I do not know. We had some slight sport throwing lances at a school of blackfish that swarm along with the ship sometime. Wherever one would come within dart on either side of the ship he was sure to get the cold steel through his hide.

The sun rose up with peaceful light
Upon the Tristine Ground
Adorning with its cheerful beam
The ocean all around
No sooner have we made all sail
Than "mast heads" is sung out.
Two up the foremast quickly jump
And two more up the Main
So high upon the crosstrees there
They seem to dangle in the air
But heedless of this awful height
They scan the sea with all their might
The crew about the decks are working
Some here some there and some a shirking
Forward the wooley Doctors galley
The rendezvous where loafers sally
When suddenly a cry is heard
only this one single word
Blows! When every eye upon the strain
turns where the spokesman looks
Blows! then blows! soon comes again
The Captain in a trice jumps up
And calls out where away—
Two points off the weather bow
And "there goes flukes" begins the row
We up two points and brace the yards

We'll see how close to wind our bark
Can crawl when need requires such work
Blows, there blows again is heard
Where away? The old man cries
Dead ahead the mast head calls
Boys, clear the boats—then clews brailed
Sends each man to his post
A minute more and "lower away"
Gives willing crews alert. The ship
is luffed, the main yard hauled aback
Three boats are in the water dropped
Men row on out the chase begun
And now we hope to see some fun
I see, he cries the red blood rise
In columns from his blow hole
Now!—sailors for the sake of home
Haul him and get the prize
We draw up near
His side is clear—
The mate has thrown the lance—
Deep buried in his oily hide
He feels that steel so often tried
And forthwith lifts his mighty flukes
Stern! Stern! My boys, stern for your lives
He'll blow his last now by St. Ives.

We backward send our trusty boat
But he is on us still
Pull two—ease three—give way—stern all
Our boat with that big tail he'll maul—
Lay off—lay off—or we'll be stove
And then our whale is lost by Jove
These orders were no sooner said
Than they were heeded and obeyed—

This poem, written in pencil, cannot be identified so may be Weir's attempt at verse. The reference to the "wooley Doctor" suggests that the cook, William Stevens, was Black.

Friday, April 4th

Another beautiful day with very light breezes, but our good ship glides along rather fast having the wind right astern, and fore-topmast studdingsail set. While making passages from one whaling ground to another, the old man cracks on all sails, in order to lose no time. Our course now is southeast by south for rounding the cape of Good Hope.

This morning being my watch below, I am engaged in making a bed quilt of calico and strips of blanket in preparation for cold and comfortless times that we expect off the Cape. This afternoon we pecked, pounded and tarred the anchors merely to keep the hands from idleness. In the evening, the captain deals out some tobacco—the best Albany oak leaf can't compare with it—I got four pounds.

Sunday, April 6th

Beautiful day—not much wind steering full and by with all sail set—quite in contrast to last night's watches, for we had rain and the wind was variable and baffling, had to brace round, trim and square repeatedly. We often have very dull and unpleasant times on deck at night, and for that reason all like boats crew watches the best as there will never be more than four hours for crack boats crew to stand but in sea watches, every other night one watch has eight hours on deck and there is a chance for three or four hours below every day. These watches we stand only when carrying sail.

With the wind changing direction frequently, the crew was kept busy adjusting the sails, using the lines called braces to swing the yards, trimming the fore-and-aft sails by easing or taking in on the lines called sheets, and sometimes adjusting the course to remain square to the wind.

Monday, April 7th

I have commenced now to try and keep some sort of a journal or log book, as the sailors call it, in which it is my intention to place the remarkable incidents occurring during the voyage should such occurrences happen.

I can't always find the chance to write when I would wish to in a watch below, for it is often impossible. Before we have taken many hundred barrels of sperm oil, I shall no doubt have something interesting to jot down—which I shall enjoy or regret as the case may be at some far distant period. Our latitude today is 31°7'S Long, 5°30'W heading southeast with wind astern at present.

Now that he is writing his journal in the form of a logbook, Weir begins to observe the sea day, which runs from noon to noon.

Tuesday, April 8th

A beautiful day after a most unpleasant night with little or no wind. We could not wish for better days than this on land, but on shipboard it is vice versa—everything is changed except the heart and no doubt that is very often.

Wednesday, April 9th

Set taut the main, main-topgallant, main topmast and royal stays—steering full and by with a stiff breeze and all sail set. I know what a beautiful sight it is to see a vessel skimming the water with all sail set; often have I watched them (years gone by) from our parlour windows, it was a great pleasure for me then, and what would I not give to be in sight of those scenes of my childhood at this present time. When I think of the disgrace I have brought upon myself I imagine I am best off here, tossing about on the boundless ocean—at least it must be so for the present. This forenoon we lowered and struck at a sunfish, but the iron striking upon the hard part of the fish's back would not enter. The shank of the iron bent double and the edge of the head taken off as though the harpoon had been thrown against a rock.

With the wind astern, the crew tightened the stays supporting the mainmast to counter the pressure of the wind on the rig. Weir recalls seeing cargo sloops and schooners sailing up and down the Hudson past West Point. The ocean sunfish, or *Mola mola*, is among the largest bony fish, with a truncated body weighing up to a ton and reaching a height of 10 feet. They have elongated dorsal and pectoral fins but only a lumpy pseudotail in place of a caudal fin. Sunfish are generally sighted when they are basking on the surface, resting on one side with the dorsal fin out of water. They seem to live in coastal waters, including off South Africa. As Weir suggests, they have very thick skin.

Thursday, April 10th

Started the sewing society again, stitch, stitch patch on patch is all the rage, here are half the ship's crew below going at it hammer and tong with their needles. Here is where I am learning famous lessons in economy; with all sorts of trades, cobblering, barbarizing, washing, tailoring, with probably many others that do tailoring, and with probably many others that do not occur to my mind at present. A whaler might well be called—Jack at all trades—for this is a little of every imaginable thing done on board a whale-ship.

It was Friday the third day of August 1855 that I left Cold Spring for the last time. To go to sea was the last thought that entered my head that morning, but how little did I know myself. I think I myself shall be better acquainted on my return should God see fit to allow it.

Friday, April 11th

Lat 33°2' S, Long 3°50' E

Sailing close hauled with a six knot breeze, sail in sight at 7:30 AM. When I had just taken my station at the wheel by 9 o'clock, we were nigh enough and spoke her—proved to be the *Nightingale* of Boston homeward bound from Shanghai. She was steering north-northwest and we southeast. What a beautiful sight this vessel presented bearing down before the wind—with skysails and top-gallant studding sails set and four jibs numbering more than 26 sails. A perfect cloud of canvas and her hull was a fine model clipper. I amused myself while pacing the deck last night in thinking over all my lady friends, I wonder if that one whom I looked upon as my star still shines for me?

What musical fevers I once had with Mr. Angel, how often have I jumped from our piazza, violin in hand, to play hour after hour with Mr. A?— I wonder if he enjoyed it half as much as I did—he would sometimes take the flute, but generally the guitar—can I ever see such times again?

How often I think of Miss Fraser and all the family. Their house has indeed been a home to me—nor shall I forget all their kindnesses. I wish Alex was with me here, I would care for nothing more. No—oh! no—it is too bad to wish him to enjoy? the misery I do—he is much better off than I am. We have been friends too long and too true to forget one another, I must write to him at the very first opportunity, and stand the chance of sending the letter home in some homeward bound vessel—should we meet one. Quite an era—(we did not scrub decks this evening). All sail set with staysails and starboard fore-top-mast-studding sail.

The 1,060-ton, 185-foot clipper ship *Nightingale* was launched in 1851 and proved to be a very fast vessel. When sighted by the *Clara Bell*, the *Nightingale* was engaged in a triangular trade route, sailing from New York to Melbourne, Australia, with passengers and merchandise, then to the coast of China, returning to London or New York with a cargo of Chinese tea. Among his "lady friends," Weir mentions "one upon whom I looked as my star," which may refer to Anna Chadwick. Mr. Angel was either a music instructor or simply a skilled musician at West Point. Miss Fraser and Alex Fraser have not been identified.

Sunday, April 13th

Had quite a severe squall last night—with considerable rain. Kept out the main topgallant sail, though it was risky—expected the order to clew up every moment—though everything cracked, we kept on all the rags she could well bear. Men were stationed by the haulliard ready to drop, clew up and clew down in a trice.

Weir's spelling for halyard—haulliard—explains the term. A halyard is the line hauled on by the crew to raise a yard or sail in the process of setting sail. The lower corners of a square sail and the free corner of a triangular jib are called clews. To set a square sail, lines called sheets draw the clews down to the yard below. To furl a square sail, clewlines are hauled to raise the clews up, leaving the sail hanging loose for furling.

Wednesday, April 16th

Last Monday the steward slaughtered the old sow "Cleopatra," our last memento of Tristan D'Cunha Island, and all hands gloried in the deed, for she had often made our watches on deck anything but pleasant. The evening of this same day, Mr. Barker the mate struck a large porpoise, from the martingale guys, which we hauled on deck, stripped of his blubber and hung upon the Main stay in a short time. Porpoise flesh is considered by some a delicacy, we eat it for a change. It tastes very much like veal but is not so firm and is of a dark color. The oil tried from the blubber is used in the binnacle lamps as it gives a remarkably clear light while burning. What with porpoise meat and fresh pork, we have lived rather high for the past 30 hours, especially while at mast head. Today I commenced to make a cap of a variety of pieces of cloth, but now I shall not have much difficulty in finishing it, for the steward has given me a sizeable piece of cloth thanks to his generosity.

In his journal, Weir labeled the pig as "Cleopatra."

How often do I think of my dear sisters while engaged upon all this kind of work? Wouldn't they laugh and criticize? But they may yet find out I am not so poor a knight of the needle after all.

Livestock, including cattle, hogs, and chickens, was one form of food that whaleships might pick up at Tristan da Cunha or St. Helena. They were generally tied or penned on deck until they were butchered. If the sailors were not using the term "porpoise" for a species of dolphin, this was probably a spectacled porpoise, a species found in far-southern latitudes. They have a rounded nose, a triangular dorsal fin, and a white belly and black back. They may grow to seven feet in length and perhaps 200 pounds. Because of its remote habitat, little is known about the spectacled porpoise.

Thursday, April 17th

Mr. Barker lowered for another sunfish, but we lost sight of him shortly after the boat left the ship. I would like very much if we could get one—the oil from the liver is said to be excellent for rheumatism and is used for many medicinal purposes.

Friday, April 18th

Scudding like a seabird before the wind with all the square sails set, our bearing at present is about Lat 37°12' S, Long 16° or nearly. At our present rate of sailing we'll soon be right off the south point of the Cape.

It is now 4 PM, the wind has increased so much that the fore-top sail was double reefed, and when a huge sea lifts us our good bark seems fairly to fly—so lightly does she float—any way if she don't fly the spray does.

The starboard anchor worked loose this morning, and we were obliged to lash it passing the rope through the hawse holes. The starboard boat is storm rigged for we are on the larboard tack and the lee side occasionally dips. By 5 PM the mainsail was furled, and both topsails close reefed. *Clara* staggers like a drunken man, you are tossed here, there and everywhere, like a ball in a box; with the pots, pans, spoons, chests etc.—you cannot resist dancing a jig. On deck it is touch and go, the seas washing over continually keep the decks so slippery that it is dangerous moving about, for they are inclined nearly fourty degrees every few seconds. Life lines have been rigged aft and the watch on deck is obliged to stop there and keep out of danger. When the watch on deck has been relieved by those from below—the general rush that is made for the forecastle scuttle is ludicrous enough—ten to one, the whole watch will be crowded about the scuttle way and a big sea will wash over—there'll be some noise heard then, and more wet jackets.

To storm rig the boats in their davits, they were lashed on their sides, with bottom out, to shed the seas. With the vessel on the larboard tack, the wind was coming from the larboard (left) side, and the vessel heeled to starboard under the force of the wind.

Saturday, April 19th

Still moving like a thing of life before the wind—A little after 11 PM wind sifted forward our beam, bringing some squalls. The mate has requested me to write in a journal for him. I have taken the job though it may be time thrown away upon such as he, however it may make me be more industrious and so be of some service.

Sunday, April 20th

Eight months from home today. Lat 38°9' S, Long 19° W Just past the extreme point of the Cape, and upon the verge between the Indian and Atlantic Oceans—We are fast moving to the scene of operation, the whaling ground, with wonderful expectations as to filling the ship in one season. One sail in sight this afternoon, bound for Atlantic shores. Another sail this morning off our weather bow on the same tack as we.

Monday, April 21st

Overhauled the helm gear, from 5 PM through the night had a smacking breeze. Lat 37° 51' S, Long 23° 43' E Many and many a weary league am I now from those I love and O what a life for me to lead among an ungodly set of men, where there is nothing but coarse and immoral language used. It is something I can never get accustomed to—and yet I have to put up with it for at least two years more.

But I am undergoing a rather rigid schooling, which cannot soon be forgotten. I regret being secluded from society, and such refinement as I have formally been accustomed to—far beyond everything, I must improve my time in every possible way—mentally, morally and physically.

Could I but hear from home and know just what feelings are entertained towards me for my past misconduct—I might be pretty well satisfied—but no! I am denied all communication—for there is no means of sending anything to me. Thus must I remain till the time of my exile has expired and then—Oh! Horrible—I may be a changed being, and even not be recognized by old friends. I shall strive hard to keep from all temptations to make myself worthy of such a father as I have. This is certainly a calamity I have brought upon myself, therefore I deserve it all.

Overhauling the helm gear means changing or adjusting the lines connecting the wheel to the tiller that moved the rudder. The tiller rope was wound around a horizontal drum attached to the wheel, then the ends were run through blocks on either side of the deck and back to the tiller. Turning the wheel tightened one end of the tiller rope and loosened the other, pulling the tiller and moving the rudder.

Tuesday, April 22nd

Wind from the east, tacked ship this morning at 6 o'clock. Heading southeast by south at noon running under double reefed topsails. Wind increased to a gale, by sunset we were staggering under a foresail and close reefed main topsail. The seas were pretty heavy, and continually washed our decks. Everything fore and aft was well soaked and washed with salt water, several seas pitched over us scarcely sprinkling the decks. The order of the day at such times is turn in wet and turn out smoking—the wardrobes in our forecastle are not very extensive, and should we have rain three days in succession, all the dry goods in the forecastle would be wet.

Wednesday, April 23rd

During the midnight watch the wind lulled rapidly and the seas were not quite so bad, and by four bells (2 AM) all sail was set including stay sails. By 7 AM all light sails had to be taken in—the seas toss the ship about most unmercifully, and everything is comfortless as usual in such weather. Steering east-southeast true course east by north on account of variation of compass, currents etc.

Thursday, April 24th

It rained in the morning—by 10 o'clock very fair and almost a calm. At noon strong breeze off the beam, on the increase and hauling to the westward. Took in all light sails by 1:30 PM. By 2 PM wind is very strong and right astern—furled the spanker, gaff topsail and jib. Steering east wind southwest. We now set the starboard fo'top mast stud'sail—but it was no sooner hoisted home, than the boom snapped off short up to the yardarm, and fell into the water towing by the sheet. Down the sail had to come and the old man coming on deck ordered a block to be hung on the end of the fore yard arm through which we rove the studding sail halliards and set the sail alow. The seas were now growing higher and higher till 6 PM when the studing sail and main topgallant sail was taken in, and now commenced the gale. By 7 PM we were scudding along with incredible speed, full 14 knots, amidst a sheet of foam and a deluge of rain, it seems madness to rush along so, but there was no alternative. We were now under close reefed main top sail and a bellying foresail—what would a landsman think to see us now—the sea running awfully high.

For a moment we would be lifted upon the summit of a great sea, amidst a cloud of spray, anon we were dropped between two mountaining heaps, where it was almost a calm, and all our headway apparently gone.

Look up—see that towering billow astern—we shall be engulphed—but no—the mass of water would rush on, our good bark would mount high upon its top, and again give a plunge and start ahead like a frightened deer.

"Entering Indian Ocean" shows the *Clara Bell* running with heavy following seas under double-reefed main topsail, reefed fore course, and a storm staysail.

Occasionally during the night, the gale would lull a little and our good bark would then be plunging and rolling at a fearful rate; the seas rushing by tons upon our decks. It was worth one's life to let go a hand grasp from the rigging for a moment—Again the gale would come upon us with redoubled fury—making the sea look like driven snow and lifting showers of water from the crests of the waves which fell upon us like heavy rain. One may imagine a sailor's life is a cheerless one in such times, and so it is to some—but not so to me, for I love to be on deck and watch the sea and sky and hear the Almighty's voice in the storm. It makes us feel that we are actually in the great presence of the Omnipotent. It serves to remind all that God is still there and watching over them. The sea is His and He made it.

How little we think so—and yet we know that His slightest thought could send us all to destruction—could send the whole universe. I should think sailors above all men should be Christians, because they seem to live in these depths of danger and God is eternally saving them from destruction. It is trying to be separated from the rest of the world—a mere speck upon the bosom of the ocean, nothing to be seen around but sea and sky.

What an atom is a ship at sea compared with the Universe, and yet God is there.

With the vessel running downwind, the fore-and-aft sails were irrelevant and were taken in. The studding sails were set below booms that were run out from the yardarm (end) of the topsail and topgallant sails, with a similar boom at the foot. When the boom broke off at the topsail yard, they suspended the sail from a block on the fore course yardarm.

Friday, April 25th

At 11 PM the gale abated a little, but before daylight a hailstorm came on, accompanied by vivid flashes of lightening and loud peals of thunder. By noon bowling along under double reefed topsails mainsail jib and spanker. Steering east as we have been doing since last night at 7 o'clock, fresh breeze from the north and west.

Thursday, May 1st

This month will be a memorable one at home on account of Emma's marriage which was set for this time—and a right gay time they were to have. But here am I—far-far away in the Indian Ocean many a weary mile from those I love—a wanderer from home—an exile by my own thoughtlessness. Our sight was enlivened just at daybreak this morning by the appearance of two sails. We showed the log and compared latitude and longitude Lat 20°57' S, Long 45°40' E. One of these vessels was English and the other French—both under royals and studding sails.

Emma Weir married Thomas Lincoln Casey on May 12, 1856.

Friday, May 2nd

One sail in sight at noon, we discovered her to be an English merchantman bound for the Atlantic; at sunset shortened sail to double reefed topsails, heading northeast steering full and by.

Sunday, May 4th

Yesterday (3rd) spoke and gammed with the bark *Isabella* New Bedford, 8 months from home, been into port about six weeks ago in Toula Bay Madagascar. They sent about two dozen terrapins on board of us, and now every turn one takes about the deck he is afoul of them.

This morning we raised what the whalemen call a stinker—a large dead whale—bore down for it and found it to be a fin back whale too long dead to be of good. The sharks appeared to hold a meeting of some kind over the carcass, doubtless in regard to its removal. The birds were not more backward in their addresses.

Commenced standing boats-crew watches last night; a sure sign that we are on the whaling ground. For the past two days we have been cruising in sight of Madagascar. This is that far famed sperm whale ground Fort Dauphin. The weather is boisterous and the coming three months are said to be typhoon season off here. Where we shall go now, I cannot say—I would not grumble to go home.

The 315-ton bark *Isabella* of New Bedford, commanded by Captain J. Lyon, departed for the Pacific whaling grounds on September 4, 1855, and returned on May 18, 1859, with 963 barrels of sperm whale oil, 1,250 barrels of right whale oil, and 6,200 pounds of whalebone (baleen), having sent home 35,000 pounds of baleen. A "stinker" was a dead whale found afloat that could be stripped of its blubber. The decaying carcass had a bad smell.

Wednesday, May 7th

Gammed with the *Almira* for about the seventh time since we crossed the line. Sent us a boat load of terrapins, but whether all hands shall indulge in terrapin soup I can't say. For the past two or three days we have been setting up the fore rigging stays etc. Almost beginning to forget what we sailed for.

Crossed the line means crossed the equator. Whalemen commonly gathered turtles, or terrapins, on remote islands and kept them on board for food. "Setting up" means tightening, which was done by taking up on the rope lanyard that ran between the pair of round, three-holed wooden deadeyes at the foot of each shroud and some stays.

Friday, May 9th

I feel that I am now beginning to break down in all good resolutions, my feelings are fast turning to disgust. Oh! For the refinements of home, sweet home—one day among dear old friends would make a new man of me. Could I only get a letter—no matter if it contained only a single line—it would be worth heaps of gold to me in my present deplorable situation. We live worse than felons and certainly we are driven as bad. The only difference is in our consciences, and truly all of us cannot brag of a clear conscience and all are well aware that many men who go whaling are perfectly destitute of the remotest conscientious feelings. My pride, though I have held it under mighty subjection, is still far from being wrecked. I am sometimes almost bursting with it, and I am afraid I shall shortly burst out in a grand explosion. At the present, impudence is a virtue, for those who have a good stock of impudence always get on the best. I have exercised it some and find it is the principal virtue here. I don't refer to insulting impudence—but cool, calculating impudence used with considerable sauce of discretion. Oh! humbug—it always gives me the blues to think of the life I am obliged to lead—all my refinements and politeness (if I ever had any) is fast scattering to the four winds of heaven. I shall never *never* go to sea again, unless it may be while traveling. Though I love the sea, there are too many loose and worthless men who live upon it. I mean in the whaling service—for I know some noble men in the Navy and Merchant service too. But the men here, their very touch is contaminating, their slightest presence pollution. At least so I feel it here.

Sunday, May 11th

No whales for us yet. Long 48°15' E Steering to the northward to try our luck off the Mahé banks till the weather is better down here.

Sunday, May 18th

We have been running to the northward and eastward for the past week, and our latitude must be somewhere between 8 and 10 South and longitude, over 50—truly a long ways from home. For the past fortnight we have had very disagreeable weather, and for the past two days we have had a great many rain squalls, accompanied with thunder and lightning. It rained in torrents all this morning watch from 2 till 7 o'clock. My expectations that were once so exalted, are now fast falling below zero. Here have we been eight months—aye—nine months from home, and have not even seen a sperm whale blow. We scan the waters as far as the eye can reach from sunrise to sunset, but see nothing, save those marine animals we don't wish.

As for right whales I don't want to cruise for them anymore. I prefer warm latitudes, for cold weather is anything but agreeable at sea. While

"Going on to a Whale, Sperm Whales" shows a boat entering a pod of sperm whales under sail, with oars peaked and the crew wielding paddles. As Weir depicts, a whaleboat would approach whales under sail when the wind permitted. The light spruce or cedar mast was set up by passing the end through a hole in the second thwart and into a socket (step) in the bottom of the boat. The sail was spread by a long, light spar called a sprit, which stood diagonally, with the upper end in a grommet in the peak of the sail and the lower end in an eye splice in a short line low on the mast. The stern oarsman tended the sheet (controlling line) under orders of the boatheader. Once the harpoon was darted, the bow oarsman disengaged the sprit, rolled the sail around the mast and sprit, and lashed it with the sheet. As the whale line ran out past his legs, he lifted the mast out of the thwart and passed it aft, where it was wedged under the aftermost thwart to hang out of the way, over the stern.

off Tristan D'Cunha I suffered more than I care to again. If we don't get a good share of sperm I suppose the old man will run to the southern ground for more whale oil. Lat 11°13' S, Long 53° E getting warm.

Sunday, May 25th

Saw a school of sperm whales last Thursday. Killed two, one of which sunk. The other *turned up* a little more than 25 bbls. Lowered again today, but without success—our boat was right upon the whale, and though I saw both the irons thrown in the finest possible way, we did not get fast. I am confident the first iron entered, but the second though directed straight enough lacked force enough to make it enter. There were about 30 or 40 whales in this school, all of a good size. But I must not omit an account of our first sperm whale. A few minutes before 8 o'clock Thursday morning, we discovered a school of cows and calves off our lee bow—about 3 miles distant—we were then steering north. We immediately bore off and ran almost due east—head on for the square heads—gradually we neared them, and new life seemed to possess us all by this time—the anticipation of capturing some of these mammals and having glorious sport. The boats were quickly in readiness and no sooner had the captain given the order to "lower away the boats" than all three boats with their respective crews were in the water quicker than Jarvis. And now our sail is set and as we are to the windward of the whales it is necessary to observe the strictest silence—each one has peaked his oar and grasped the paddle—and we all tug away with a hearty good will, but the whales seem to anticipate danger and have quickened their speed.

After chasing for nearly two hours, Mr. Perry's boat got fast and a few minutes after Mr. Welch's boat was fast securely to a cow—our boat was a half a mile off at this time but we shortly pulled up—so soon as the other boats got fast we took in our sail, stowed away the paddles and bent to our oars with desperation—soon we were close by the waist boat's whale and now the mate sung out to Johnson our boatsteerer "stand up and give it to him" and he did give it to solid. An instant and the water is white with foam, but the whale doesn't stop to fight—off he darts dragging our two boats humming through the water, shortly he slackens his speed. The mate is now in the bows, lance in hand, "haul line" he shouts—we soon haul close, and the mate has darted the lance chock to the socket in leviathan's very vitals—again the water foams, but this time it is red with the blood of the whale and off he starts swifter than ever—the line hums around the loggerhead and the water has to be freely hove upon it. The sparks fly—look out there! Foul line the tub man cries out, and quicker than thought, a whole bunch of line is whisked past everyone in the boat. Another bunch flies out, this time dislodging several of the crew from their places. The line is now displaced from the chock and drags across the mate's body. Soon the whale slackens again, and we reset the line. We again haul up and keep close to the whale the mate lances him three or four times more. Oh! Look out—he is moving his circle, he'll be fin up shortly—look to your oars men—keep clear of him while he's in his flurry. Now the water foams and flies, stern off for your lives, we quickly stern off and look again—the whale is dead he is *fin up*. If all sperm whales are as easily captured as this, we won't have much difficulty. It is far different from right whaling, for they fight like mad with their enormous flukes. But a square head doesn't lift his flukes so threateningly. When this whale lay alongside before cutting in, I had some leisure to survey him. My first impression was that it was a half formed work of the creator, but when I thought about who made that creature I could see beauty in every inch of it—Oh how wonderful and curious are all thy works oh thou Creator.

Lore suggests that Nantucket whalemen first encountered sperm whales around 1712, although the animals had been known for centuries. Sperm whaling began in earnest from Nantucket in the 1730s. The quality of the oil, and the waxy material called spermaceti contained in the "case" in the sperm whale's forehead that could be molded into fine candles, quickly made the sperm whale the principal target for New England whalemen. The sperm whale (*Physeter macrocephalus*) is the largest toothed whale, having about 40 conical teeth in the lower jaw that fit into sockets in the upper jaw. Sperm whales largely feed on giant squid in the depths of the seas. They can dive more than a mile deep and remain underwater for more than an hour. Sperm whales communicate and echolocate by producing loud clicks that travel miles through the sea. The long forehead and case of spermaceti are thought to help with echolocation. With their low-set eyes, it has been suggested that they swim upside down while hunting squid in the depths. Sperm whales live in pods of between six and 20 individuals made up of females and juveniles. When a male matures at about age 18, it leaves the pod to swim alone in higher latitudes, only returning to female pods to mate. Sperm whales seem to have a natural lifespan of about 70 years, and a mature "bull" sperm whale may reach 50 feet in length and 100,000 pounds in weight. Females may grow to 36 feet and 30,000 pounds. When their digestive tract is irritated by the horny beaks of the giant squid, they produce a waxy material called ambergris, which was highly valued as a fixative for scent in perfume. Agile and very intelligent, sperm whales fought with their flukes and their jaws when attacked.

Friday, June 6th

Dropped anchor this morning at 10 o'clock off Port Victoria—Mahé Island, one of the Seychelles. The fruit here is abundant and as cheap as sand—I have already had some delicious bananas and plantains. Oh how I wish I could send some home—I shall be obliged to eat for those at home and for myself too. The land breeze is delicious, and we will soon have liberty given us to go ashore, then how shall I feel, like a freed bird—I feel it coming on already.

Mahé, Seychelles, is located about 1,000 miles north of Madagascar. First visited by Europeans in the 1700s, it was named for Bertrand-François Mahé de la Bourdonnais, a French governor of Mauritius. The British seized Mahé from France in 1812 and made it a colony. About 60 square miles in area, Mahé rises to almost 3,000 feet above sea level. The main port, Victoria, on the east coast, is sheltered by a band of small islands.

Sunday, June 15th

We have been reveling in all sorts of tropical fruits—bananas and plantains may be had almost for the asking for a Marké (a coin valued at one cent and a half), you can buy a bunch containing 20 or 30—a treat for a dozen. All kinds of fruit can be obtained remarkably cheap and there are a great many different kinds and I have eaten as much as 20—half of which I do not know the name of. The "coeur de boeuf" is a magnificent fruit about the size of a man's fist, and tastes like "blanc mange" highly flavored with vanilla, the seeds resemble the castor bean, though hardly as large and seem to grow miscellaneously scattered in the fruit. When this fruit is ripe and fit to eat, it is so mellow and soft that to let one fall from the hand to the ground would immediately transform it into a squash. This morning the Portuguese have gone ashore till noon, and the Yankees will go this afternoon. The weather begins to get unpleasant—frequent showers of rain come sweeping upon us from the hills.

There are two barks here in company with us, the *Montgomery* and *Eugenia*. They were both here before us. Captain Cottle of the *Eugenia* is sick and ashore, he has a good crew and they all like him. They regard Sunday in the right light on board there—make no sail, stand no mastheads—nor do any kind of work, while with us it is considered a good day for getting whales, perhaps the difference between Sunday and any other day with us is this—every six days of the week we break out the hold and the seventh day we don't, but are hauling and pulling on ropes, trimming sails, setting and taking in sails, etc. etc. etc.

The Seychelles grew many fruits that were unfamiliar to Americans, including passion fruit, star fruit, African custard apple, and jaboticaba, as well as breadfruit. Weir seems to describe African custard apple (*Annona senegalensis*). The 248-ton bark *Montgomery* of New Bedford, commanded by Captain William B. Chapman, departed for the Pacific whaling grounds on August 23, 1855, and returned on June 18, 1858, with 385 barrels of sperm whale oil and one barrel of right whale oil. The third mate and his boat's crew were lost, fast to a whale, on November 19, 1856. The 356-ton bark *Eugenia* of New Bedford, commanded by Captain William Cottle, departed for the Pacific whaling grounds on November 6, 1855, and returned on May 17, 1859, with 1,351 barrels of sperm whale oil and 215 barrels of right whale oil. William Cottle of Tisbury, Martha's Vineyard, was third mate of the bark *Popmunett* at age 22 in 1843. He was second mate of the ship *Menkar* in 1845, first mate of the ship *Navy* in 1848, and master of the ship *Cambria* in 1851, followed by the *Eugenia* in 1855.

Thursday, June 19th

Villainous weather, head to the south and west at 5:30 PM, took in sail and luffed to the wind heading northeast by north. By morning steering west. Last Tuesday the 17th, we weighed anchor from Mahé and not without regrets from all. Each watch had three and a half days of liberty—not half as enough but we must try and content ourselves—we are nothing but slaves.

While strolling about among the hills I continually thought of my home and all the dear ones that I left there. I wonder where they imagine I am now. Oh, how I wish I could hear from them. I'd give anything.

Oh, the Island here, coconuts can be had for the trouble of picking them, and sometime the trees are so low that there is not much difficulty in getting them. After walking about in the heat of the day, it is delightful to sit under the shade of the trees and drink the milk from fresh picked coconuts. The same fruits that we can have fresh here taste totally different when obtained at home. Sugar cane suffered extensively—some of the boys have stowed a quantity in their berths. Every edible upon the island seems well adapted to supply the wants of those living in hot climates—they are all so remarkably cool and juicy. Pineapples seem to grow all over the place.

We took on board ten pigs large and small, a dozen or more of fowl, 200 coconuts, any quantity of bananas and plantains, with small quantities of numerous other delicious fruits. We also stowed ten cords of wood and 20 casks of water (120 bbls) so that our vessel is now filled to the hatches and in good sailing trim.

Monday, June 23rd

At 3 PM sail in sight off the lee bow, bore down and spoke her between 4 and 5 PM—proved to be the bark *Iorco,* merchantman of Salem homeward bound from Zanzibar Island. About 5:30 PM we discovered a whale ship off the lee beam, ran off for her and at dark set a light, but she would return no signal. A few minutes past 10, beginning of the midnight watch, she was seen within half a mile of us, we wore ship and double reefed the topsails, and again lost sight of her. At midnight the bolt rope of the jib parted with a loud noise tearing the sail to ribbons. Steering west with fresh breeze from the south Lat 4°36' S, Long 43°42' E.

The *Iroco* of Salem cannot be identified. Possibly Weir refers to the bark *Iosco* of Salem, which sailed to Zanzibar in 1855. The small archipelago of Zanzibar lies along the African coast northwest of Mozambique. After a period of Portuguese control, Zanzibar came under rule of the Arabic Sultan of Muscat and Oman. Zanzibar traders dominated the trade in African elephant ivory, gold, enslaved Africans, as well as the cinnamon, cloves, nutmeg, and pepper of the islands themselves. To wear ship is to change tacks by steering away from the wind, bringing the wind from one side of the stern to the other, whereas to tack ship is to steer through the wind, bringing it from one side of the bow to the other. A bolt rope is a heavy line sewn around the edges of a sail to reinforce it.

Tuesday, June 24th

At 10 AM we raised what we supposed to be sperm whales. Lowered the boats at 10:30 AM and chased till 1 PM without a particle of success—steering northwest.

Wednesday, July 2nd

Land in sight the coast of Zanguebar, at 4 PM took in the light sails, and tacked ship, running on the starboard tack, wind still from the south heading southeast by south, crossed the line last night Lat 00° Long 42°E

"Zanguebar" is an earlier spelling of Zanzibar.

Friday, July 4th

Where are the days of my childhood, those happy innocent feelings of joy that once filled my breast? Can they ever return? Can I ever again feel so free and light of heart? No never—never shall I enjoy such a life again, this rash act has spoiled my future existence—the constant thought of what I have done must continually be uppermost in my mind, to frustrate all my happiness.

But this I richly deserve. How could I be so inconsiderate, so unkind as to leave my dear father in this manner? Has he forgiven me—can he ever forgive me? Oh! for one kind word from him, what potency it would contain—and must I wait two more weary years, before I shall see or hear of him. Still I have hopes of getting word in some way—though there is so much uncertainty. But why do I write thus? Why? Truly I cannot help it, for these and similar feelings are continually bearing upon my mind. How could it be otherwise after such ungenerous and ignoble conduct? I feel as though I deserved to be discarded forever from the good will of those at home.

Well! to proceed, this day has been one of a little enjoyment with us—coconuts, roast pig, minced pies, soft tack, ginger cake, pepper sauce, molasses, pepper, rice and pickles, was our bill of fare—quite extensive for sailors wound up the day by during salutes with a couple of packs of fire crackers and a grand concertino given by the steward and myself on an old tin pan and a cracked flute. My home, my home, my happy home—can I ever enjoy a home again? Lat 1°34', Long 45°45' E

Captain Robbins allowed the crew to celebrate the 4th of July in style, including roast pig and soft bread. Again, Weir showed his musical side.

Sunday, July 13th

Lat 2°34' S, Long 42°31' E

We are now cruising off the eastern coast of Zanguebar. Here have we been battling ever since leaving Mahé—without the slightest particle of success. The weather has been unpleasant all the while, with a very strong current from the south. Have been standing sea watches since last Thursday in order to carry more sail at night and beat to the southward.

Sunday, July 20th

Lat 2°20' S, Long 29°14' E

Steering for the Mahé banks, Seychelle Islands—course southeast by south. Eleven months out this day, and all this time not a word from home. I hope they have heard from me before this.

Friday, July [25th]

Bird Island in sight, lying off and on this afternoon, the jolly boat and bow boat sent ashore after eggs. At noon we brought a full boat load on board as we could obtain them without the slightest difficulty. The island has an area of about a square mile—it is almost barren, a few bushes and many weeds constitute all the verdure—but the great feature is the myriads of birds and eggs that are upon it. The birds hover about the island like a dense cloud, and it is necessary to walk carefully else you will crush eggs at every step; and it was something of a job to drive the birds out of the way, in order to get their eggs.

Bird Island is the northernmost island of the Seychelles, about 60 miles from Mahé. This coral island was named for its many birds, including a very large population of sooty terns. It was also home to tortoises.

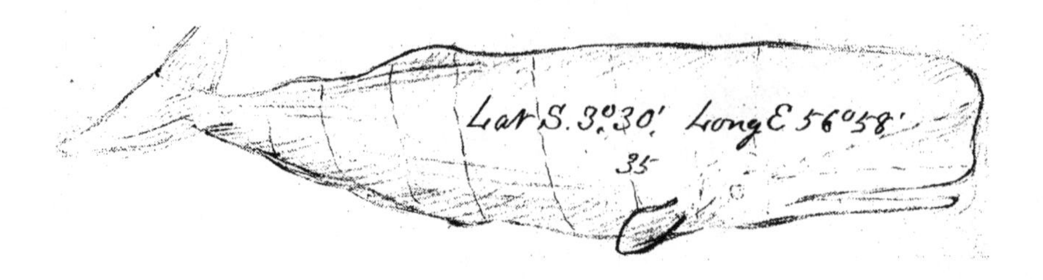

Saturday, July 26th

Lowered all three boats this morning after square heads, the starboard boat was the only successful one—they got a cow which will make about 35 bbls. The other boats had a hard chase and came aboard between 12 and 1 PM. Hard at work all this afternoon cutting in, finished by 6 o'clock and then wore ship—and now we have got a dirty job for tomorrow—to lean the blubber and cut horse pieces. It will be hard work to know what day it is.

> "Square head" was the whaleman's term for the sperm whale as in profile it has a long, horizontal "forehead" and a vertical "brow."

Friday, August 1st

Lat 3°27' S, Long 56°17' E

At sunset took in light sail, hauled up the courses heading southeast by south.

Just one year agone today, I saw the last of that fairy spot West Point. Full two years more must pass ere I can hope to see it again—and what change may take place during that time—whatever they may be, I must know they are for our good.

But I would like to get home again—Saw a very large sword fish this morning, that deadly enemy of the sperm whale, that is excepting ourselves and other whalemen. We often see "devilfish" or *diamond* fish as they call them on board here, they put one very much in mind of a mammoth bat or vampire with horns and tusks which look very white. We have never caught any of these but I should think the average length of one would be six feet. They are ugly looking creatures, and no doubt they could do some mischief, when fastened to. The shovel-nosed shark is

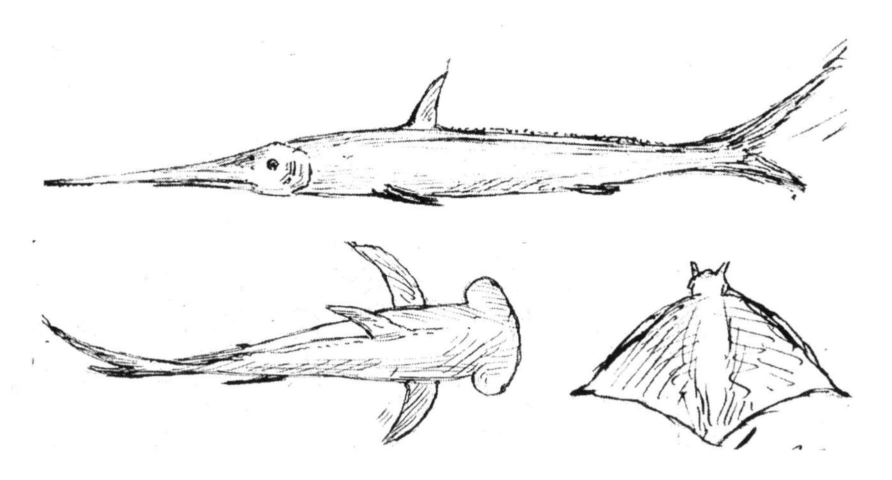

another curious fish, said to be perfectly harmless though it has a loathsome appearance. The largest kind of shark is the bone shark, he lives upon the same kind of food as the right whale, having bone in the mouth, though it does not grow very long and is poor stuff. These fish are hunted for the oil which their liver contains, their size is that of a 20 or 25 bbl whale. They are harmless, and when struck always sound and lay sulkily on bottom from which they must be hauled and then they are killed without difficulty.

The Indian Ocean was home to six species of "billfish," including three species of marlin, the shortbill spearfish, the sailfish, and the swordfish (*Xiphias*). Swordfish can grow to 10 feet in length and 1,000 pounds in weight and can swim at up to 50 miles an hour. They consume a wide range of fish but are not known predators of the much larger sperm whale. By "devil fish," Weir probably means a manta ray. By "shovel-nose shark," Weir refers to what we call the hammerhead shark. In the Indian Ocean, the great hammerhead (growing to as much as 20 feet) and the scalloped hammerhead (growing to about 13 feet) are the most likely species sighted. "Bone shark" is another name for the basking shark, which lives in temperate waters, typically grows to more than 20 feet in length, and is a slow-swimming filter-feeder with a very large mouth for taking in plankton.

Sunday, August 3rd

One year ago, I left Cold Spring to try my fortune on the wide world. I have left home and happiness, dear relations and friends. I left my dear father—the thought maddens me—has he forgiven me? Does he still look upon me as his son? Oh! for a few words from him, how relieved I should feel.

But my bosom friend Morris—Lou Morris—O why was he not spared? Had he lived, I know I should never have been here. How often had I told him all my heart, of my intentions, and how he used to persuade and assist me to stay and finish my course at the Foundry. Can I ever find another, such a good and kind friend and brother, it is impossible, he was truly a father to me—he exerted everything to assist and encourage me, and now he has gone. Your memory my loved friend is sacredly kept. The whole family though I am but little acquainted with them have showed me much kindness—never can I forget them. Mr. Richard Morris, his father, is truly a noble man, his sisters are of the highest stamp, so refined and gentle. It is impossible to forget them.

But I cannot forget my situation, can they ever receive me at home, as formerly? or is my name blotted from their happy list? O God! I do wrong to think this of those who loved me so dearly—my father, brothers and sisters. And does she think of me with scorn? No—I shall not think so.

Here Weir alludes to but does not fully explain the event that he considered so disgraceful that he had to run away to sea. His good friend Lewis Morris died a few months before Weir left the West Point Foundry at Cold Spring. Weir refers to "she" here, but the context is not clear. He may refer to his favorite sister, Emma, or he may refer to his future wife, Anna Chadwick, whom he had probably met several years earlier.

Sunday, August 10th

Lat 4°35' S, Long 42°40' E

Time glides along swiftly—perhaps the wind carries him away from us! However he never goes too swiftly with me, knowing every week brings me so much nearer my return home. How little did I feel the true meaning of those dear songs so often sung at home, "Our way across the sea," "Be kind to the loved ones at home," "Do they miss me at home" and several others full as truthful and touching.

At this present time and in my present situation I find the purest relief to be alone and sing or think of these songs and everything that I used to sing with my dear sister Emma; she is always in my thoughts, and it is my earnest hope and prayer that you are contented and happy, my dear Em. And a blessing to your new home—one peep at you all would give me new life. Louise I know is happy for she cannot be otherwise with Seymour. Do they often think of me, how I used to bother them? I wonder what Walter can be doing, as he knows more of my circumstances than many others at home. He doubtless does not look so hard upon my actions. The longer I remain here so much the more deeply do I see into my past life. I must concentrate all my energies on my arrival home, and strive to live as a Christian; to live a useful life, and to be a blessing to my dear father—not a curse. As yet I dare not dream of happiness and contentment, above everything keep me from the sea. I love friends and relations and home too much to desire separation.

"Our Way Across the Sea" has not been identified. The song "Be Kind to the Loved Ones at Home" was written by I. B. Woodbury and published in 1847. The poem "Do They Miss Me at Home," by Caroline Atherton Mason, was put to music by S. M. Grannis and published in 1852. With its theme of homesickness, it became extremely popular. Weir refers to his sister Emma, with whom he was closest, their older sister Louise, married to Truman Seymour, and the eldest of the siblings, Walter.

Sunday, August 31st

Little or no breeze heading southwest by west this morning, heading southwest by south Lat 2°56' S, Long 56°11' E. For the past three or four weeks we have been cruising on and about the equator. The coast of Africa has been in sight several times (Zanguebar) and while anywhere within 4 degrees of land we have enjoyed miserable weather—rough, with a considerable sprinkling of rain squalls. Our greatest distance north of the equator was about 40 miles. We are now steering towards the Seychelle and Almiranta Islands—at all events that is the rumor among us. Not a square head have we seen since our last capture, perhaps the old man is going in quest of better luck. For the past ten days we have stood sea watches, in order to carry full sail during the nights and make a quick passage to the Mahé ground. I have not noticed the sea so quiet as it has been today, since we left home—generally when it is calm there is a swell on, making the sails flap with an unpleasant noise against the masts—this weather I do not like—give me a stiff breeze and I am comfortable.

What would I not give to be at prayers in the painting room tonight, to hear my dear father pray. Oh! how thankful I am that he has brought us up in such a way, the thought of it has restrained me from many an unpleasant deed and whenever I read the family prayers I hear my father's voice—this has been the greatest connecting to him. And oh! how often do I feel that he is praying for me—I must deserve his love, I must strive to be a son to him.

Here Weir confirms that the family would gather in Robert W. Weir's studio for prayers each evening.

Sunday, September 7th

Miserable weather, under very short sail, heading southwest or nearly so. At 7:30 AM set full topsails, stiff breeze and rugged seas. Aldabra Island (in sight) wore ship at 6:30 PM and 1 AM.

Aldabra is one of the Comen Islands, and a right pleasant sight it was to see land, though it is but two months ago that we saw any land so near (four miles off). This island covers about eight square miles, and is very low—there is plenty of verdure upon it but no fruit.

Whalers sometimes put in here to get wood and send their boats crews fishing along shore. During the past four or five days we have had quite a gale for this latitude 9°S—have been under double reefed topsails about all the time, and one night we "lay to," with close reefed main topsail fore spencer and staysail. Today I have been writing a little in a letter to Walter. If I could only hear something about the good folk at home, I would consider myself a lucky boy, but I almost fear this happiness would be too great for me—will be denied me on that account—still I hope and so I shall do to the end of the voyage.

I wonder if my darling little pet Charlie ever thinks of me. I wish she could always remain the same artless girl, and then by the time I arrived home, I would have many a curious tale to tell her. I could wish to tell her stories forever. I spend many an hour of my solitude pleasantly in thinking of her and everybody and everything connected with her—but this thought will occur, in spite of all, she may be greatly changed when I meet her again.

Part of the Seychelles group of islands, Aldabra is a large coral atoll 700 miles southwest of Mahé and about 225 miles northeast of the Comoro Islands, which are located at the head of the Mozambique Channel, between the African continent and the northern point of Madagascar. "Charlie" seems to be a young woman friend from West Point with whom Weir had grown up, but her identity has not been determined.

Saturday, September 13th

Lat 11°39' S

Comoro Island in sight. Fresh breeze from the south, variable, one sail in sight. At sunset shortened sail, at 7 PM tacked ship heading west-southwest, 9 PM set the light sails. This morning breeze freshening. 6 o'clock tacked heading east-southeast.

Sunday, September 14th

Time wears apace. Here we are trying to weather "Comoro" and run for Johanna where it is the skippers intention to anchor and repair and paint ship, preparatory to a cruise off Fort Dauphin. Comoro has been in sight since last Monday, now we would be within five or six miles of it, then a calm would come, and the current take us 20 miles from it. It is very

strange, when we have been 50 or 60 miles from this island it could be distinctly seen a little before sunrise, but when the true sun was about one degree above the horizon, not a vestige of land could be noticed—sometimes it would be visible at noon, and always just at sunset. At night we can see the light of the crater, there being a volcanic peak on this island. Johanna is about 95 miles to the south and west from us, and if we proceed no faster than we have done for the past week, we shall probably reach it in the course of a month or six weeks. Such is fate—calms, currents and winds oppose our progress, our feelings rise and fall with the winds—a seven-knot breeze makes us cheerful, a dead calm sober and growly—a gale puts us in ecstasy.

By Comoro, Weir probably means Grande Comore, now Ngazidja, the westernmost of the Comoro Islands. Running north to south, it has two volcanic peaks, the higher one standing 7,746 feet above sea level.

Tuesday, September 16th

By 2 PM breeze is freshening, at 4 PM tried to speak and gammed with the brig *Corops,* Captain—hard to say. Late from Mozambique for Mohilla. Loaded with fruit and slaves. Our captain boarded them and enjoyed a highly edifying conversation with this Arabian captain. He offered them money for fruit, but they would take none, and placed about two bushels of delicious oranges in the boat, which the captain divided among us all when he returned—we have never tasted anything as good and refreshing.

Wind from the southeast very light and variable—heading south-southeast or thereabouts. At midnight, the wind shifted to the east and north, blowing fresh. At 3:30 AM, morning watch, in shoal water on the Comoro banks—could see bottom plain, made five and seven fathoms. By 5 AM the wind died away.

Mohilla is Mohéli, the smallest of the Comoro Islands. Muslim seafarers established a trade in enslaved Africans before the eleventh century. By the 1800s, an "Arabian" slaving vessel was most likely carrying Bantu captives from the slave market in Zanzibar to ports on the Persian Gulf. It is estimated that they carried about 3,700 enslaved Africans per year in the 1800s. By Comoro banks, Weir may refer to the Vailheu Bank, a coral bank just west of the westernmost island of Grand Comore.

Saturday, September 20th

Lat 11°53' S, Long 42°28' E

Light breeze heading south by west, "Comoro" still in sight to show our slow progress. Caught a *Dolphin* with the grains this forenoon, the color of this far famed fish is certainly beautiful and while dying, all the hues of the rainbow vibrate upon its skin. The form is rather stiff, being about five times deeper from back to belly than from side to side. We have not seen many before during the voyage, and this is the first that has been caught. They are a top water fish and eat about as well as others of the class.

"Dolphin" can refer either to a marine mammal or to a species of semitropical fish now called mahi-mahi. Here Weir means the fish. Noted for the colors its scales gave off as the fish died, the dolphin fish was valued as food on board ship—for the officers when only one was taken; for the crew if more were available. Dolphins were usually speared with the multi-pronged fish spear called a grains, often by an officer standing on the chain bobstay under the bowsprit.

Thursday, September 25th

Here we are within ten or a dozen miles of Johanna. The cable was hauled up day before yesterday, anchor cleared and everything ready to let go at a moment's warning. But these abominable calms and currents will thwart many a wise man's intentions. For the past week, Mohilla, Comoro, Mayotta and Johanna have been in sight. Comoro is about 60 miles from Johanna and can be seen from there at sunrise, noonday and sunset in good weather. These calms try our patience severely. "Hope deferred maketh the heart sick" here it is illustrated. Our bodies are so much in need of something fresh, that we can relish nothing that we have on board—bread and molasses seems to be the staple food of all. Oh for a few of those delightful fruits with which the island abounds, a few months full would serve to put new nature in our veins.

Friday, September 26th

Passed Saddle Island (a small Island close to Johanna) at about 3 PM when the wind died away, so we had to down boats and tow in. At 4:30 PM the anchor was fast in the mud off "Johanna" and the Arabs are now swarming about the ship offering shells, sugarcane, dates, milk and bananas for tobacco. They are a curious race, but their appearance seems familiar from having seen such correct representations of them in pictures. How much dear father would like to be here and go among them. What happiness it would be for him to see the natives in their houses—their picturesque costume would soon get his pencil in motion. This is a sickly place and the Captain intends on leaving as soon as possible, though each watch will have an occasional run on shore. There is not a great variety of shells to be met with here, but I must trade for a few to remember the place by—perhaps Emma would like some.

Weir depicts a Muslim man at Johanna (Anjouan) in the Comoro Islands. He wears a loose thawb, with his keffiyeh wrapped as a loose turban, and he appears to carry a scimitar, perhaps as a symbol of status.

Johanna Island, now known as Anjouan, is a volcanic island, the easternmost of the Comoro Islands. Triangular in shape, with peaks rising to almost 5,000 feet, it has a large bay on the north side where the main settlement was located. Originally settled by Africans and Indonesians, by the 1800s it was ruled by Hadharem Arabs from present-day Yemen.

Wednesday, October 1st

Poor weather, painting ship and receiving taro. Captain Robbins sold the jolly boat for about 200 dollars' worth of taro and potatoes. Have been ashore considerable since we anchored and I find plenty to amuse myself with—Prince Selim Houssan has been my guide, a boy about 16 years old, quite bright and intelligent. He showed me all about the town, and when I asked to see some Arab women he said suppose you kill me I show them; however I did see one, though I had to enter the house by force—and the Arab whose harem we entered made a great to do about it—beat his head, pulled his beard and uttered all sorts of incoherent sounds that soon collected a crowd, from which we vamoosed without much ceremony. On entering some of the houses where the women had been employed, we saw some of the most beautiful embroideries that no doubt they had hastily left on our entrance—this was in Abdalla the king's house.

Taro roots or corms, a diet staple from east Africa to the Pacific islands, could be treated by whaleship cooks much like potatoes, boiled, baked, or fried. As a devout Christian, Weir did not understand, or did not respect, Muslim religious restrictions on interaction with women of the faith. Nevertheless, his young royal guide offered the hospitality expected of devout Muslims.

Thursday, October 2nd

Been on the lookout for a fair wind all night. At sunrise weighed the mud hook with a stiff breeze from all the northern points of the compass. 1 PM steering south. The south side of this island is formed like a horseshoe and it was in this bay we anchored. We saw plenty of fruit upon the island that was not ripe—mangoes, pineapples, figs, etc. etc. Took on board several bushels of tamarinds, which the boys are now putting up in molasses—being an excellent drink in warm climates. Some of the coconut and palm trees upon Johanna were of enormous height, at least 100 feet high. The mango tree was the most beautiful of all, such thick foliage.

The north side of the island has a wide, horseshoe bay. There is a smaller bay on the southwest side, but with no settlement.

Thursday, October 9th

Fresh breeze from the south heading southeast at sunset. Tacked heading southwest set sea watches and kept all sail out. At midnight tacked ship, heading southeast by east. Raised sperm whales a little after 6 this morning, kept off for them. 7:30 AM boats were lowered and we chased hard till 11 AM about which time the waist boat got fast to and killed a small fellow.

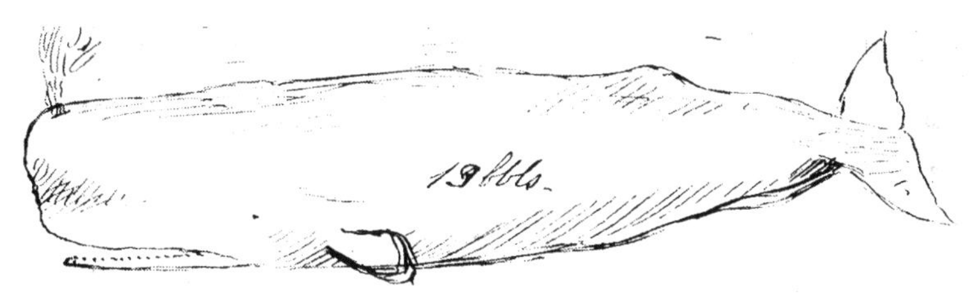

Weir copied this entry onto the wrong page and crossed it out. It is put here in the proper sequence.

Sunday, October 12th

Since our departure from the "Comoros" we have been beating down the "Mozambique Channel." Lat now is 16°43' S, Long 41° E. Last Thursday we got into a large school of sperm whales, and after considerable chasing, the waist boat succeeded in getting a small one. This fellow stowed down about 13 bbls, but there was a vast amount of spermaceti in the oil.

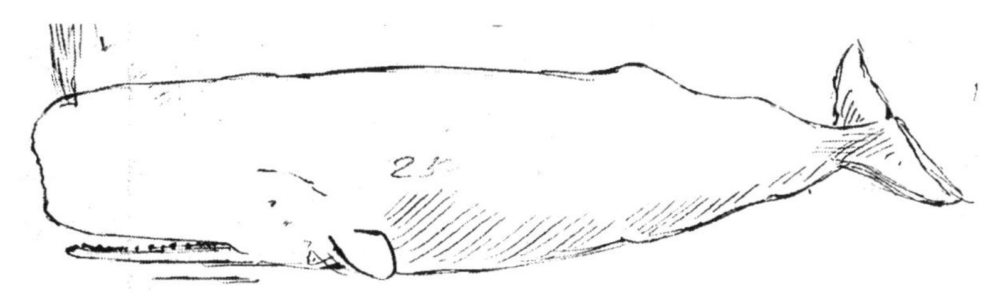

The Mozambique Channel is the waterway between Mozambique on the west coast of Africa and the island of Madagascar. It is 1,100 miles long and 260 miles wide at its narrowest point. The Comoros lie at the north end of the channel, and the Mozambique Current runs south through the channel. Spermaceti is the waxy material contained in the case in the sperm whale's forehead and may serve a function in the whale's click communication and echolocation. When exposed to air, spermaceti cools and solidifies and may later be processed and molded into very clean, bright-burning candles.

Tuesday, October 14th

Moon almost totally eclipsed at 2 AM. Commenced at about 12:30 and ended at 3:30. We had a good view of it for the sky was clear—wonder if they will know of this at home! Home! Home.

Thursday, October 17th

Our boat succeeded in getting a 25 bbls whale last Thursday out of a very large school. Raised them breaching Wednesday afternoon, chased all night, lowered at half past 6 AM. Fourteen months out and still in debt. On my return home I shall get quite as destitute as I was on the day I left. This is a hard life, and I shall be happy enough to get out of it, though everyone says it will be almost impossible, and that I will go to sea again. I think I can overcome this infatuation—I shall try my utmost to do so. The excitement is what I shall miss, but the love of friends and country ought to conquer this. I have felt the loss of them once—again—I never care to. Lat 19°6', Long 47° Mozambique Channel.

By debt, Weir refers to the fact that he owed the cost of his outfit, plus the items he had purchased on board, and they had not yet taken enough whale oil to pay his expenses, given his "long lay" or share. Experienced seafarers in the crew expected that Weir would eventually find the adventure of seafaring irresistible and would return to sea. Weir disagreed, but he would find his resolve weaker than he thought.

Sunday, October 26th

We make but slow progress against this northerly current; calms have prevailed the greater part of last week, so that we have made but a few miles to the southward. Today we have a fresh breeze from the west and south, with every rag set—the wind is variable and orders are continually given to trim yards etc.—sea getting rugged.

I commenced a letter to sister Em last week, but it may be a long while before I shall get the chance to send it.

This afternoon at 4:15, Johnson a noble young man and friend of all was lost. There was a heavy sea on, at 2 PM the martingale guys gave way, and had to be immediately repaired for the loss of our bowsprit and fore-topmast would have followed. A little after 4 o'clock Johnson was serving the guys together just abaft the martingale, when the mate sung out to the captain telling him all was right. We braced forward the main yards and as the ship's head came to the wind she met three heavy seas, causing her to pitch heavily—twice Johnson was dashed several feet beneath the water, but he still kept his hold, against the awful pressure. The third time he was madly forced under and the great power of the waters was too much for him—he was dragged from his hold—ropes, and every available thing were hove to him, a boat was cut away and manned, but all too late. The startling cry of man overboard thrilled all to their very hearts' core—what could be done?

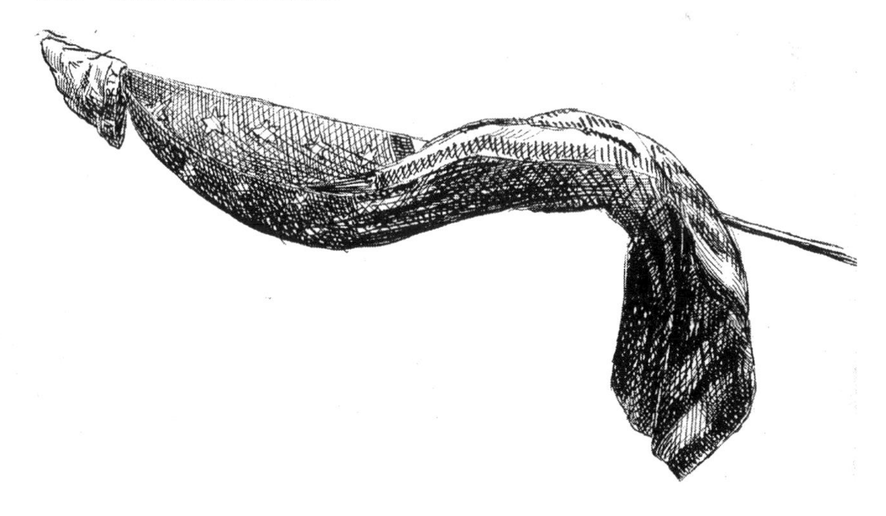

In such weather the seas running so high, the ship dashing madly through the water. Oh! what feeling filled my breast as I rushed to clear the boat and help save him who was overboard in such a gale; I thought of my wronged parent, of my present situation, the many sins upon my heart. Oh! what happiness it must be to be a Christian, and always be ready to die. I have taken this event in its true light as a warning for us all. O! that I could think more of God; so merciful to all sinful mortals. Oh! God give me that strength to love and fear Thee eternally.

This sad event cast a deep gloom over all our company—for William Johnson was a friend of all—he was a man of good principles to think of his poor mother, brothers, and sisters is truly melancholy, for he was their main dependence. Often had we conversed together of home and all dear recollections of home, had even planned for the future and now he is taken away. God in His mercy has seen fit to do this, it is well.

The martingale is a stay running from the end of the jibboom on the bowsprit down to the dolphin striker, a vertical strut under the bowsprit. A pair of martingale guys run from the foot of the dolphin striker to each side of the bow of the vessel, and together the martingale and guys, with tension provided by the dolphin striker, counteract the upward pull of the fore-royal stay at outer the end of the jibboom. If the guys came free, the jibboom and fore-royal mast would likely break off. Johnson was finishing the repair by serving the line when he was washed off his insecure perch. Boatsteerer William C. Johnson had gone to sea as a greenhand in the whaleship *Eugenia* in 1851, then served as a boatsteerer in the *Congress II* for its 1852 voyage.

Monday, October 27th

The captain sent for me this morning and told me to take Johnson's place as boatsteerer—it is unexpected, for I did not think he had a competent man on board. I feel grateful for his kindness and shall strive to do my best for the voyage.

Perhaps Mate Barker had noticed Weir's good behavior in his boat and requested of Captain Robbins that he be promoted as his new boatsteerer.

Sunday, November 18th

Sail in sight, proved to be the *J. Dawson,* 13 months out, 100 bbls sperm. Gammed all the afternoon, at dark luffed to the wind heading north-northeast, wind east and north. At 6:30 AM land in sight—Madagascar, tacked ship heading east-southeast. Raised sperm whales at 10 o'clock this morning, lowered in company with the *John Dawson,* boats returned without success.

The 237-ton bark *John Dawson* of New Bedford, commanded by Captain Amos C. Baker, departed for the Atlantic and Indian Ocean whaling grounds on October 3, 1855, and returned on May 6, 1859, with 577 barrels of sperm whale oil.

"First Fort Dauphin Whale" possibly depicts the waist boat being "stove" by a sperm whale, which crushes the boat as the boatheader lances it and the crew leaps overboard.

Sunday, November 23rd
At 8:30 PM gammed with the bark *Massasoit* 31 months out, 700 sperm.

The 200-ton bark *Massasoit* of Mattapoisett, commanded by Captain E. B. Handy, departed for the Pacific whaling grounds on April 11, 1854, and returned on December 20, 1857, with 726 barrels of sperm whale oil, having sent home 26 barrels of sperm whale oil.

Monday, November 24th
Stiff breeze and rugged sea, three sails in sight. At 6:15 AM we raised sperm whales, lowered at about 9 AM. Waist boat got fast and was taken to windward smoking, by noon quarter boats came aboard—beating with the ship, all sail set. At 1 PM lowered the quarter boats and chased whales to leeward, waist boat sent signals of distress—Ship also—pulled hard to the rescue and found her stove. Larboard boat got fast and settled the business. By 4 PM whale alongside, and a big fellow he is too.

Sunday, November 25th
Commenced cutting in at daybreak, by noon had all in but the head. Two sails seen.

Wednesday, November 26th
Spoke the ship *Martha*, no letters for me, and she is but six months from home, but soon we hope to see other and later sails.

There were several whaleships named *Martha*, but this was probably the 301-ton ship *Martha II* of Fairhaven, Massachusetts, commanded by Captain Timothy C. Spaulding, which departed for the Indian Ocean whaling grounds on May 20, 1856, and returned on April 1, 1860, with 1,001 barrels of sperm whale oil and 189 barrels of right whale oil. Vessels often carried out letters for other whaleships they were likely to encounter on the whaling grounds.

Wednesday, December 24th

Chasing whales all day—lowered the boats four times—before breakfast, after breakfast, after dinner and supper. No success attended our movements, the weather was too good.

Monday, December 29th

Gammed with the *H. H. Crapo*, *chock a block* bound for home shortly. I sent four letters on board. One to A. V. F. with a note enclosed to Emma, another to Walter and to Father and Seymour.

The bark *H. H. Crapo* of Dartmouth, Massachusetts, commanded by Captain Archelaus Baker Jr., departed for the Indian Ocean whaling grounds on June 12, 1854. About a month after meeting with the *Clara Bell*, on January 19, 1857, the *Crapo* was lost at sea. Only Captain Baker and one sailor survived the sinking. A. V. F. has not been identified but may possibly be a relative in the Ferguson family. Again, Weir refers to his brother-in-law Truman Seymour.

1857

Tuesday, January 6th
Gammed with the small bark *Acorn* five months from home. No letters, not even for the captain. What is the matter?

The 215-ton bark *Acorn* of Provincetown, Massachusetts, commanded by Captain Puffer, departed for the Atlantic and Indian Ocean whaling grounds on July 25, 1856, and returned on June 12, 1859, with 595 barrels of sperm whale oil.

Wednesday, January 7th
Gammed with the *United States* six months out—as unfortunate as usual—have they forgotten me? I think not.

The 217-ton bark *United States* of Westport, Massachusetts, commanded by Captain A. Hicks, departed for the Indian and Pacific Ocean whaling grounds on June 23, 1856, and was wrecked on May 1, 1860, with 550 barrels of sperm whale oil on board, having sent home 175 barrels of sperm whale oil.

Sunday, January 11th

I've struck my whale—lowered at 11 AM. By 12 o'clock, Mr. Barker put me on a noble whale—took him head and head. I got up and gave it to him solid—whiz, whiz, whiz—it seemed but a moment and all the line was out of one of our tubs—160 fathoms. I hold the turn, he shortly slacks and again comes up to blow, the starboard boat gets fast and within an hour he is fin up. The waist boat was there with the bomb lance, but did not have an opportunity to use it. Poor Deacon Welch and his bomb gun get but little encouragement from the other whalemen! Keep your weather eye open for a fighter my boy, then blaze away your popping things.

"Striking Attitude" seems to show Weir in the process of harpooning his first sperm whale. After darting his first double-flued iron, he will attempt to set the second one as well. Behind him, the bow oarsman handles a paddle but will quickly drop it to secure the mast and sail. The dark post by Weir's leg is the crotch in which the harpoon handles rested until they were darted.

Weir implies that the whale immediately sounded, taking 960 feet (160 fathoms) of line after it. After Weir and Mate Barker changed ends of the boat, Weir at the stern held a turn of the line around the loggerhead to impede the whale, and it soon returned to the surface. While whaleboats commonly worked independently, on the *Clara Bell* they often collaborated in taking a whale. Christopher Brand of Norwich, Connecticut, and others developed shoulder guns for whaling, beginning in the late 1840s. These weapons, commonly called "bomb-guns" in the outfitting records, fired an explosive iron or brass projectile about an inch in diameter and from 16 to 21 inches long. The guns were heavy, weighing about 24 pounds, and had a range of about 60 feet. They were effective when whaling in the Arctic, when a whale might escape under the ice, or when hunting whales with very thick blubber. Some boatheaders found them safer than a hand lance wielded up close, but bomb lance guns occasionally exploded or otherwise injured their user.

Monday, January 12th

This morning got the case and junk on deck—enormous—talk about your elephants, mastodons and mammoth monsters of the earth. What are they compared to a noble whale—how the land folk would open their eyes to see such a head as this.

I am 21 years old this day—oh! how the time flies, and these moments can never be recalled. I wish I was somewhere near civilization. I'd feel better satisfied. To remain here 18 months more seems awful, but I've battled it so far why not finish with Gods help—if father could see me now!

The case is the cavity in the sperm whale's forehead, which contains the spermaceti. The junk is the fibrous, oily mass of tissue below the case and above the sperm whale's jaw. These were both cut away from the body and hoisted on board for processing.

Tuesday, January 20th
Saw whales at 4 PM. Lowered—larboard boat, struck and drew—poor chance. Two sails in sight. Wore ship heading east-southeast. Wore ship at 2 AM heading north-northeast, at 6 AM spoke the bark *Eugenia* with a whale alongside.

Monday, January 26th
Strong winds from the east, heading northeast. 6:30 PM wore ship heading south by east, at midnight wore round heading northeast. At daylight raised sperm whales, coming to leeward. Lowered at 7 AM without a particle of success, came on board before noon, *Eugenia* in sight. Lowered again at noon—no use, by 3 PM gave up the chase as hopeless.

Tuesday, January 27th
Still chasing whales in company with the *Eugenia*, returned on board disheartened with our ill luck. At 3:15 PM wore ship heading northeast by north, sunset wore ship southeast by east. At 10:30 this morning saw whales, lowered a little after 11 o'clock, between 12 and 1 o'clock starboard boat got fast to a pretty gay customer, turned him shortly with the help of the bomb gun. 3:30 PM had the whale safe by the flukes alongside. Got part of the jowl and blubber on deck by 8:30 PM, knocked off then for the night. 4:30 AM commenced operations again, by a little past noon had all valuable parts on deck.

Saturday, January 31st

Saw whale, and lowered without success. *Jos. Maxwell* close to—boats down—no use. The captain came on board with his boat, 18 months out 800 bbls sperm. 6 AM saw whales and lowered with our usual run of luck.

The 302-ton bark *Joseph Maxwell* of Fairhaven, Massachusetts, commanded by Captain Andrew P. Jenney, departed for the Pacific whaling grounds on November 3, 1855, and returned on August 7, 1858, with 1,495 barrels of sperm whale oil and 16 barrels of right whale oil.

Thursday, February 5th

Blowing quite a gale from the east. Under foresail and double reefed main topsail, sail in sight, signaled her—proved to be the *Swallow*, new ship lately from home. Weather quite boisterous during the night. 7 AM wore ship and furled the foresail, heading northeast by east.

The 439-ton ship *Swallow* of New Bedford, commanded by Captain Herman N. Stewart, departed for the Indian Ocean whaling grounds on October 9, 1856, and returned on December 22, 1860, with 600 barrels of sperm whale oil and 800 barrels of right whale oil.

Friday, February 6th

Still lying to. Prospects as bad as ever, considerable rain and more wind during which pleasant time we must turn to and wash ship laying the deck with no economical feelings. Wore ship at 3 PM heading south by east. At sunset *Swallow* in sight—too rough to speak or gam.

Saturday, February 7th

No improvement—heading south-southeast, plenty of rain during the night, at 7 AM wore ship and set the foresail, heading north-northeast. 11 AM blows great guns, all hand on deck to take in sail, lying to under spencer and staysail, took bow boat on deck.

Sunday, February 8th

Better times, 2:30 PM under close reefed topsails and foresail, rainy during the night. From 12 PM till daylight little or no wind set all sail, 8 AM little rainy, at noon good weather—prospect of its being bad.

Wednesday, February 11th

4:30 PM set the foresail—Fine weather, steering east, 9 PM sail close to off the lee bow, during the night had light breeze from south. 5:30 AM spoke the ship *Herald,* they saw whales at sunset last night. 7:30 AM raised sperm whales, lowered at 8:30 AM without success. *Herald* close to with four boats down, no success—quite calm.

The 303-ton ship *Herald* of New Bedford, commanded by Captain Henry H. Slocum, departed for the Indian Ocean whaling grounds on August 13, 1855, and returned on July 30, 1858, with 1,020 barrels of sperm whale oil, 282 barrels of right whale oil, and 3,000 pounds of whalebone (baleen).

Friday, February 13th

Weather improving a little, set double-reefed-fore topsail. 6:15 PM wore ship heading southeast by south, while wearing ship got into a school of sperm whales. The sea was rough, the sun was setting—of what avail was the sight. 7:30 AM raised whales going to windward like blazes, seas quite rugged 11:15 AM lowered the larboard boat, no use.

"Meridian, Our Mate" shows First Mate Barker taking a noon sight with an exaggerated quadrant to calculate the vessel's latitude.

Monday, February 16th

2 PM four sails in sight. 4 PM, raised whales while gamming with the *Maria*, lowered the boats with usual good luck. Finished the gam by 7 o'clock, during the night heading northeast, wind east and south.

The 202-ton bark *Maria* of New Bedford, commanded by Captain Joseph Abbott, departed for the Indian Ocean and Pacific whaling grounds on September 1, 1856, and returned on August 11, 1859, with 684 barrels of sperm whale oil.

Sunday, February 22nd

Rough weather, under double reefs heading east-northeast. 6 PM wore ship heading south by east, furled fore top sail, took in jib spanker and mainsail. Fort Dauphin has the true reputation of being a hard old place.

Fort-Dauphin, now Taolagnaro, on the southeast coast of Madagascar, was a small anchorage on a rugged coast. It is not clear if the *Clara Bell* landed there or just fought the wind and seas off that stretch of coast.

Monday, February 23rd

Wore ship at 4 PM heading southeast by south. 10:30 AM wore ship heading north-northeast, weather is still quite rugged. Wore round at 9 AM heading south-southeast, two sails in sight, broke out water and corn.

Tuesday, February 24th

1 PM—wore ship heading north-northeast, 6 PM wore ship heading southeast by south. Strong easterly wind. 2 AM wore ship heading northeast, two sails in sight.

Wednesday, February 25th

Wore ship at 1 PM heading south-southeast, 4:30 PM squared yards steering west by south. At sunset set sea watches, kept on all sail during the night. By daylight this morning steering west by north sent down the fore top sail for repairs, going at it hammer and tongs.

Thursday, February 26th

2:30 PM bent the foretop sail, during the night heading west-northwest, quite squally, rainy and unpleasant. 10 AM steering north, wind south.

Friday, February 27th

Blowing a pleasant gale, steering north by east and north-northeast. Land in sight at daylight, by noon chains up and anchors ready for dropping.

Saturday, February 28th

We are nearing St. Augustine Bay. The weather is beautiful. Sandy Island off the larboard beam, 3:30 PM we were safely anchored in Augustine Bay, opposite the delectable city of Tentrock, what a sight, our fancy bark is crowded with natives of both sexes—a most sickening smell of beef tallow pervades the atmosphere. The canoes about the ship are as thick as flies 'round the sugar bowl. Hope we'll get out of this soon.

St. Augustine (or Augustin) Bay is on the southwest coast of Madagascar, at the mouth of the shallow, winding Onilahy River. Captain Robbins later recalled it as a "lovely harbor." He wrote: "Entering there, we dropped our mudhook off Tent Rock. Very well named, I call that rock, for it is shaped like a tent and as white as new canvas. It is just a mile from shore." Expecting to be boarded by the natives, Robbins had all loose items carried below. "Hardly had I got upon deck again when the canoes, deeply laden with their savage freight, came splashing for us. Then there was a wild scramble up the ship's side. . . . They had grass mats around their loins. Some had arrayed themselves in dirty flannel shirts. Their women wore nondescript garments of cotton cloth, and had their hair done up in little knots like nutmegs and covered with grease." Robbins negotiated with the head men to obtain permission to land and resupply with firewood and water. "The regulation tariff was thirty yards [of cotton cloth] to the chief and five fathoms [30 feet] to each of his advisors." But Robbins also had to pay the natives who lived up the river. "I wanted to tow our raft of casks a long way up their river to get fresh water, and I was aware that we should have to wait over night before we could bring them back to the ship. We must make fast friends with the natives or they would steal the hoops off our casks" (Robbins, *The Gam*, 196–97, 198–99). The Malagasy people of Madagascar descended from Austronesian people who migrated west by canoe from present Indonesia nearly 2,000 years ago and Bantu people from Africa who arrived a few centuries later. The Mahafaly ("Those Who Make Holy") ethnic group lives along rivers in arid southwest Madagascar. More Bantu than Austronesian in origin, they were pastoralists, herding zebu cattle and fat-tailed sheep, growing rice, corn, millet, cassava, and yams, and fishing.

"Madagascar Corn Mill" shows Mahafaly people near St. Augustine Bay pounding corn in a hollowed stump. Women did most of the food preparation while the men tended the cattle and sheep and cultivated family plots. They grew cotton and wove cotton cloth, wearing it loosely wrapped as shown in Weir's sketch.

Sunday, March 8th

Weir uses the term mongoose for the long-tailed, sociable little animals he encountered, but they were actually ring-tailed lemurs, which, unlike the mongoose, are native to Madagascar. Weir seems to interpret maky, the Malagasy name for them, as macocki.

Starboard watch had liberty last Sunday and the larboard watch are ashore today. Watermelons are obtained in abundance here, also muskmelons, bananas are common, sugar cane is plentiful. Limes and lemons can be purchased at the rate of one head tobacco per bush! Shells are plentiful though it seems tobacco and cloth is low. I should like very much to get several spears, shells and different kinds of cloth, but I am minus the articles to trade with.

Bullocks, sheep and goats are plentiful and cheap, the cattle are the originals of pictures we often see, *oriental* cattle with a hump over the fore shoulder. The sheep have tails weighing 10 and 15 pounds, a mass of fat that I should think it would be unpleasant for them to lug about—they do not have wool, but hair like goats and other cattle, and they are a sheepish looking animal. There are plenty of *mongoose's* or *macocki's* here, the resemblance of which seems to be between the monkey and cat. They are brought here for trade very tame, and soon grow attached to one.

Have got all our wood and water, and tomorrow will receive a lot of pumpkins, and such pumpkins, they are about the best edible upon the island. This morning had a visit from the *Massasoit* and *Smyrna* lying in Toula [Tulear] Bay about 18 miles south of us.

Living near rivers in an otherwise arid region, the Mahafaly also cultivated melons, pumpkins, fruit, and sugarcane. Both men and women used tobacco so the sailors might give up their own stock in trade. Since the natives grew and wove cotton, they were not eager to trade for American fabric (though the head men were). The humped zebu cattle, of Southeast Asian and Indian derivation, were the most common cattle species on Madagascar. The *Clara Bell* took on several zebu bullocks. Captain Robbins wrote: "The bullocks were large and fat, with humps on their fore-shoulders. Splendid fellows they were, sleek as silk, the finest I ever saw! And the prices!—you could buy them for an old flintlock musket and a few brass tacks. . . . We would purchase those bullocks on the shore and then we would swim them off to the ship and hoist them in by their long horns" (Robbins, *The Gam*, 202). Eventually they would be butchered for fresh beef. The Malagasy fat-tailed sheep is a variety of African long-fat-tailed sheep found in southern Madagascar. The 219-ton bark *Smyrna* of New Bedford, commanded by Captain George Bliss, departed for the Indian Ocean whaling grounds on December 9, 1853, and returned on September 9, 1857, with 701 barrels of sperm whale oil. Tuléar, now Toliara, is on a small bay north of St. Augustine.

Monday, March 9th

Finishing trades and getting livestock and pumpkins on board. The decks are already strewed with terrapin, all want to leave here.

Tuesday, March 10th

Weighed anchor sometime between 5:30 and 6 AM. Very light breeze, towed off with the boats, wind soon came to our relief. The decks are covered with sheep, goats and terrapin, from the size of a piece of chalk upwards, as convenient. Everyone feels dull—and I feel ten times duller. Off for Fort Dauphin.

Tuesday, March 31st

Weather bad as usual, wore ship at sunset, weather heading south-southeast and southeast. 6 AM wore round; heading northeast by east. At 8 AM most glorious to relate we raised sperm whales, by 8:30 the starboard and waist boats were in hot chase, Mr. Perry got fast, and the mate being sick, the captain lowered in the larboard boat. Though unwell I went and enjoyed the sport as well as anybody. 10 AM this leviathan lay *fin up*, by noon had him snug alongside, and cutting falls all ready. After a hasty dinner, commenced cutting him in with a will. And while Ambrose *a Gee* was upon the whale's back adjusting the blubber hook, to heave in

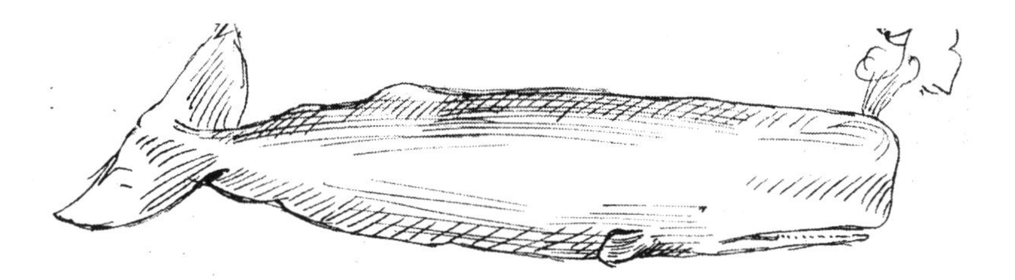

the jaw, he received a severe bite from a shark, which almost deprived him of his arm.

Normally the captain commanded the starboard boat, the first mate (with Weir as boatsteerer) commanded the larboard boat, the second mate commanded the waist boat, and the third mate commanded the bow boat. Ambrose may have joined the *Clara Bell* in the Azores as he does not appear on the original crew list. To set the blubber hook, by which pieces of the whale were hoisted on board with the cutting falls, a man with a lifeline would climb down onto the dead whale. The Indian Ocean has eight species of shark. The most likely ones to bite Ambrose were the bull shark, possibly the great hammerhead shark, the great white shark, the oceanic whitetip shark, and the tiger shark. Attracted by the scent of blood in the water, sharks would swarm to gorge on the carcass of a dead whale, sometimes in such numbers that the whalemen had to fight them off with cutting spades.

Friday, April 17th

Weather continues good, towards evening gammed with the *Eugenia* and *Messenger*. This morning raised sperm whales, ship *Messenger*'s boats hard in chase, wind increasing, whales to windward.

The 291-ton bark *Messenger* of New Bedford, commanded by Captain Isaac H. Jenney, departed for the Indian Ocean whaling grounds on August 14, 1855, and returned on March 31, 1859, with 260 barrels of sperm whale oil, 1,330 barrels of right whale oil, and 8,800 pounds of whalebone (baleen).

Saturday, April 18th

About 2 PM I was most unceremoniously roused from my berth to lower in the boat. Came on deck found the wind had shifted and was blowing great guns, brought us to windward of the whales. Our boat lowered, we set the sail and fairly jumped from sea to sea before the wind, so soon as the sail was set I took the iron in my hand, a moment and we were alongside of a noble monster, and about a ships-length off on either side were two of the *Messenger's* boats. I drove one iron home. The whale roused and shook himself, when the other iron entered his side and he was gone—whew! how the sparks flew from that loggerhead, in a few seconds our two tubs were empty (300 fathoms) and the end of the line in my hand with two turns about the loggerhead, I held on solid, the seas wash over us and the mate lifts the knife over the line, when it slackens, and the whale came up half way between our boat and the ship. The starboard boat had lowered and was now upon him. Frank gave him two irons and off he starts, this way that way and every way. Our boat was now full of water, so after hauling about 20 fathoms of line all hands had to turn to and bail boat. When this was done, we hauled in line right bravely, and both boats were side by side and almost in darting distance with the lance when he turned flukes, and though we drogued the lines he left us minus—blowing a gale.

The loggerhead is a hardwood post set in the small after deck of a whaleboat. The whale line runs from the large tubs in the bottom of the boat in which it is coiled aft to the loggerhead, where it is wrapped before running forward down the center of the boat, under a rope "kicking strap," and through a chock at the very bow before leading out to the harpoons in the whale. The turns around the loggerhead slow the line as it runs out and make the weight of the boat a drag on the swimming whale. The friction of the line could nearly ignite the loggerhead, so it and the line were repeatedly soaked with water. As Weir notes, between them the tubs held 300 fathoms, or 1,800 feet, of whale line. A fathom was a seafaring six-foot measure. A drogue was a wooden block or bucket that could be attached to the end of the whale line as a brake to tire the swimming whale and possibly allow the whalemen to retrieve the line.

Saturday, April 25th

Gammed with the bark *Julius Caesar* of New London, 6 months out, 500 bbls whale oil.

And here I met Jerry Cowles who was formerly a student with Professor Mahan; he had not much news for me, though the little he gave interested me some. Mr. Cowles was at Emma's wedding and saw father there though he did not speak with him he was looking well. Even this news gave me joy for this was nine months later than when we sailed.

The 347-ton ship *Julius Caesar* of New London, Connecticut, commanded by Captain Henry W. Bartlett, departed for the South Atlantic and Indian Ocean whaling grounds on October 11, 1856, and returned on May 28, 1859, with 311 barrels of sperm whale oil, 1,598 barrels of right whale oil, and 5,600 pounds of whalebone (baleen). Weir may refer to Jerry Sedgwick Cowles (1832–1871), son of New Yorker Jerry S. Cowles (1802–1877), who moved to Macon, Georgia, in the 1820s and became a leading proponent of railroad development in the state before returning to New York as a coal dealer. Prof. Mahan is Dennis Hart Mahan (1802–1871), an 1824 graduate of West Point who excelled in engineering and, in 1832, resigned his commission and became the Academy's professor of civil and military engineering and chairman of the engineering department. His writings and teaching on military tactics and fortifications were influential among US Army officers from the Mexican–American War through World War II.

Saturday, May 2nd

Tacked ship heading southwest. Raised sperm whales 5 PM off leeward beam, lowered with no success—quite rugged, quite dark when we came aboard, during the night heading east-southeast, 2 AM wore ship.

Monday, May 11th

Standing sea watches to make a passage. Steering east by south, breeze freshening. 3 PM took in the light sails, 5 PM double reefed the topsails and took in jib and spanker. Steering east by north, at dawn of day made all sail, 7 AM steering east-northeast. Joe Piko, a Gee, has been sick with Madagascar fever since we left Augustine Bay, and the general opinion is that captain is taking him to Mauritius.

Joe, or Jose, was likely a native of the Azorean island of Pico. The exact meaning of Madagascar Fever is unclear. Numerous mosquito-borne viruses that induced fever in humans thrived in the tropical climate, including malaria, dengue—sometimes called breakbone fever—and perhaps even rift valley fever. Mauritius lies about 500 miles east of Madagascar.

Wednesday, May 20th

Beautiful weather, [Bambous] in sight, moderate breeze from the south and west, steering west.

Bambous is the tallest mountain on Mauritius, standing 1,832 feet above sea level.

Wednesday, May [27th]

Anchors down off Port Louis Mauritius, captain ashore all night, washed ship this forenoon and sent Joe ashore.

The island of Mauritius was settled before the year 1000 by Arab mariners. France took control in 1715, and Britain seized the island in 1810. Heavily cultivated with sugar, Mauritius had a slave-based economy until the 1830s. Thereafter, indentured servants from India worked the fields. Port Louis, the principal city, lies on the northwest coast.

Saturday, May [30th]

Captain came on board at 5:30 PM and we weighed anchor, steering west and north, standing sea watches.

Thursday, June 11th

Bound for Mahé. Just finished taring down. 4 PM gammed with the *Pioneer* three years out, 900 whale 100 sperm. Finished the gam by 8:30 and such another dirty craft we have not seen this voyage, during night head east-northeast.

Tarring down means waterproofing the standing rigging with an application of pine tar. The 231-ton bark *Pioneer* of New Bedford, commanded by Captain Thomas F. Lambert, departed for the Indian Ocean whaling grounds on June 17, 1854, and returned on April 9, 1858, with 389 barrels of sperm whale oil, 801 barrels of right whale oil, and 6,000 pounds of whalebone (baleen).

Saturday, June 20th

3 PM Bird Island in sight, 4 o'clock sent a couple of boats ashore for eggs, returned 6:30 PM well loaded. Steering north till 9 PM, luffed to the wind heading east-southeast, 3 AM hauled aback fore yards.

Wednesday, June 24th

Land in sight heading southwest by west, strong wind from south and east at sunset, double reefed the topsails and tacked ship heading east by south. At 12 o'clock midnight watch tacked heading southwest. Daylight cracked on all sail steering for Mahé—anchor cock billed, cables bent, and all things ready to come to anchor.

Noon, dropped the mudhook off Port Victoria Mahé, abreast of our old friend *Eugenia*, four days in. While here all the chains and iron work about the ship was scraped and painted with coal tar, and it was daubed on with no sparing hand. We had liberty watch and watch from the day we anchored till leaving, with the exception of about three days for kicking up a row.

Bananas and plantains have suffered as usual where there is plenty and of the best growth. Coconuts are looked upon with contempt all other edibles had to take it.

A cockbilled anchor is hoisted off the anchor deck and suspended from the short, heavy timber called the cathead. With the anchor cable attached, it can be quickly dropped from the cathead. Coal tar was a by-product of the cooking of coal to produce coke or coal gas. In addition to medicinal uses, it was a preservative for wood and a rust-preventative for ironwork. Weir seems to indicate that the watches had liberty on shore on alternate days, although they were confined to the *Clara Bell* for about three days after fighting or otherwise misbehaving on shore.

Saturday, July 4th

All hands ashore on the 4th of July, formed a procession, the idea of which was to cheer different houses and individuals to get treated. *Chain lightning* flashed some, this morning (4th) the bark *Hector*/Capt. Johnson and wife arrived and anchored some way outside the reef. Sunday evening a few of our boys delivered summary justice to the mate of the *Hector*, merely an illustration of what is yet to come with others.

The 380-ton ship *Hector* of New Bedford departed for the Pacific whaling grounds on November 17, 1856, and returned on July 19, 1860, with 894 barrels of sperm whale oil, having sent home 910 barrels of sperm whale oil. Captain Johnson was another master who brought his wife to sea. Apparently, the crew of the *Clara Bell* became rowdy and attacked the mate of the *Hector* while on shore.

Monday, July 6th

All hands on board—two in the rigging, two flogged, several put in irons and the old boy to pay for nothing.

Weir glosses over the act of insubordination in assaulting the *Hector*'s mate and possibly causing damage ashore, for which Captain Robbins refused to pay. To instill punishment, Captain Robbins sent two men aloft, put irons (handcuffs) on several others, possibly confining them in the hold, and flogged the two worst offenders. A traditional punishment at sea, flogging was whipping the back of a man tied to the shrouds ("seized up"). In the navy, a cat-o'-nine-tails with nine cords was the usual instrument. Flogging was outlawed in the US Navy in 1850, but it was still practiced by some whaling captains. They generally used a single strand of small line and wielded it more to humiliate than to injure.

Thursday, July 10th

Money and liberty stopped, plenty to do and nothing to be seen that has been done or will be done.

Sunday, July 13th

Weighed anchor and went to Ladig where we dropped anchor, took in wood and recruited with fruits and livestock.

Weir seems to refer to the island of La Digue in the Seychelles, a granite island ringed with coral reefs about 35 miles northeast of Mahé. It was settled by the French in 1789 and later became home to liberated enslaved people.

Monday, July 14th

Weighed anchor from Ladig and went to Praslin where we left Andre our pilot, now we are off for whales again, hope for some luck.

Praslin is the second largest of the Seychelles and lies about four miles northwest of La Digue. Andre was a Seychelles pilot hired to help navigate among the islands and avoid the reefs.

Friday, July 17th
Fresh breeze from southeast heading east by north, still under double reefs. At 6 PM wore ship heading southwest. At 11 AM raised whales, or rather a whale and came near running him down with the ship, no use to lower.

Friday, July 24th
Off Dennis Island

1 PM the lee clew of the old main top sail gave way; 2 PM bent another top sail and at the same time whales were raised lowered before three and the starboard boat fastened. 5:30, half an hour before sunset, took out their two tubs of line and part of our boats first tub, at sunset the waist boat opened her battery with good effect, causing the enemy to cave in shortly. By dark the monster was fin up, all this time our crew was hauling line like good fellows, sent a lantern from the ship, and by 8 PM had Mr. Whale safe alongside our trusty bark. Daylight this morning commenced cutting in, by noon had all the body and jaw on deck—passable weather. Wouldn't it be gay if the folks at home could see me now, up to our knees in gurry and grease, the tryworks blazing away, and smoking the men who work in the waist almost to suffocation—they look like so many devils.

Here I must stand with the second mate driving away at the fires, piling on the scraps, pitching the minced blubber in the pots, skimming scrap and fritters and bailing boiling oil—oh! It is delightful? Abominable, horrible—here am I toasting before a fire of whale scraps, but those at home may be melting with the heat of the sun.

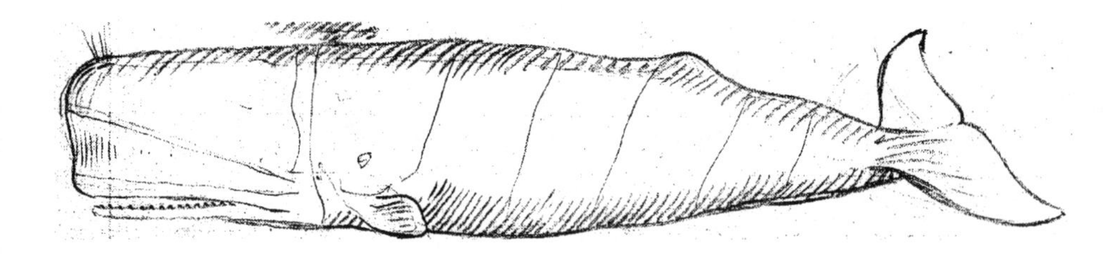

Weir refers to Denis Island, a small coral island 37 miles north of Mahé and near Bird Island in the Seychelles. A clew is the lower corner of a sail. Lee clew of the topsail means the lower corner on the side away from the wind. As Weir describes, his boat had tied its whale line to the end of the starboard boat's two lines before Second Mate Welch used the shoulder gun to kill the whale, which they towed to the *Clara Bell* in the dark with a lantern to show their position. After cutting in, Weir worked with the second mate at the forward side of the tryworks, putting the minced "bible leaves" of blubber into the pots, skimming off the flesh scraps after the oil was rendered, pitching scraps into the brick furnaces below the trypots to burn, and bailing out the boiled oil into a copper cooling tank alongside the tryworks. The oily black smoke from the tryworks filled the waist—the midships portion of the deck—making it very uncomfortable to work there cutting and mincing blubber.

Monday, August 3rd
Comoros

Chasing whales, all three boats came near fastening several times, 6 PM boats returned crews well used up, took in the light sail. 9 PM tacked heading southwest. At midnight got in shoal water, hauled aback head yards and fished till 3 AM, braced forward on starboard tack heading east, at 8 AM double reefed the topsails, wind blowing quite fresh.

In shallow water, with a chance to catch fish, the vessel was put aback, with mainmast sails filled and foremast sails braced against the wind, to hold its position while they fished for three hours.

Monday, August 24th
Near Johanna, sail in sight, kept off for her, found our old friend *Eugenia* bound in—a pleasant surprise. 4 PM tacked heading east-southeast, Comoro, Mohilla and Johanna plainly seen, tacked about a dozen times

"Selim Ahib" seems to depict Selim Houssin, Weir's guide at Johanna (Anjouan) in the Comoro Islands. He wears a taqiyah prayer cap. Weir may have misunderstood *sahib*, meaning owner or mister in Arabic, as *ahib*. He also shows an older man, who appears to wear a tarboosh or fez, and a woman in a burqa. Referring to the local culture and clothing, Weir remarked, "How Father would enjoy himself among these people."

during the night. This morning *Eugenia* way off to leeward. At noon our mudhook was fast off Johanna and the usual number of Arabs swarming our decks. 4 PM the *Eugenia* came in with a smacking breeze, while here we took in water, taro, potatoes etc. etc.

Daubed the bends with coal tar, got taken in and done for by the wily Arabs. Each watch had dinner at Selews *Hotel.* Rambled about on shore with Prince Selim Houssin for a guide, at night returned tired out. How Father would enjoy himself among these people.

Again, tacking means changing direction by turning the bow through the wind, which required more coordinated bracing of the sails than did wearing (turning the stern through the wind). The bends was a thick strip of perhaps three planks located along the hull's widest point, between the waterline and the planksheer at deck level. The bends both strengthened the hull and served as a rubbing strake. As the bends was commonly unpainted, pine or coal tar was applied as a wood preservative. The vessel was now back at Johanna (Anjouan) in the Comoros Islands, southwest of the Seychelles.

Saturday, September 12th

Mozambique Channel

We are now pretty well advanced towards the centre of the Mozambique Channel. Have been in company with the *Eugenia* since we weighed anchor from Johanna. Have enjoyed quite pleasant weather, and beautiful moonlight nights, keeping all sail set day and night, gam about once every three days.

Last night steering south by west moderate wind from east-southeast *Eugenia* off our lee beam, crew engaged in making bunt gaskets.

Bunt gaskets are lines attached to the yards to secure a furled sail once it was rolled into a "bunt." With the bunt rolled up on top of the yard, the gaskets were wrapped tightly around sail and yard and tied off. When the sail was set, the gaskets were untied and coiled.

Sunday, September 13th

At 1 PM Saw the *Eugenia* keep off, presently saw sperm whales three miles off our lee bow, with *Eugenia*'s boats in chase. Got our boats ready immediately if not sooner, and by 1:45 were hard in chase, and gaining rapidly on the *Eugenia*'s four boats. At 4 PM Mr. Perry and Mr. Welch got fast solid, and at the moment I stood up to be in readiness for a 35 bbls cow, the second mate of the *Eugenia* fastened to a whale nearby and our cow bolted, however we went to Mr. Welch's assistance and in the meantime Mr. Perry left his cow *fin up* and succeeded in killing a 15 bbl calf. Had our three whales alongside by 6:30 PM, *Eugenia*'s boats all out of sight to lee ward.

5:30 AM commenced *cutting in*; by 7 o'clock had 30 bbls of blubber on deck, went to breakfast. *Eugenia*'s boats came alongside and reported what they had done among the shoal, fastened to two whales and saved one; as the two ships had mated we must divide the game, so Captain Cottle took away our calf—not without considerable grumbling on our side. 10:30 AM had all our blubber on deck and all hands pretty deeply engaged in the greasy work of getting the blubber cut up into horse pieces for mincing. *Eugenia* close to. We are now under full topsails, heading southeast and south-southeast, wind south-southwest.

Sometimes, when hunting in proximity, whaleships might agree to "mate," working in collaboration and sharing equally in the take of each vessel.

Monday, September 14th

Cutting and mincing the blubber like blazes—pretty work for the Sabbath. Started the fires at 5 PM, people at home have just about gone to church. Oh I wish I could be with them, but there is no use in growling, t'won't bring me home any sooner. The whale tries out well—at least his oily hide does. *Eugenia* about one mile off our lee quarter, all sail set.

Made all sail at sunset, steering south, wind east by south, this forenoon wind variable, at 12 noon tacked ship, heading south-southeast.

Saturday, September 19th

Under close reefed main top-sail and foresail, blowing a gale, raining in torrents. Now I think we can appreciate the value of a home: let there be ever so much discontent, there are still some true hearts, ever ready to sympathize with us in whatever difficulties our frail natures may have brought us. Oh! That my Father may yet spare me to be again united in that family circle.

Sunday, November 1st

Dull times, dull times, no whales but for all that plenty to do, every day the same dull tune is played, oh for the exciting cry—There blows—to raise an excitement, to lower and chase whales even if we do not succeed in getting any would relieve the monotony of this abominable life on shipboard.

Friday, November 6th

Lowered for whales this morning and Mr. Perry was fortunate enough to get one. While *cutting in* spoke the *Elisha Dunbar* 900 bbls, homeward bound.

The 257-ton bark *Elisha Dunbar* of New Bedford, commanded by Captain James L. Lincoln, departed for the Indian Ocean whaling grounds on November 14, 1854, and returned on March 26, 1858, with 902 barrels of sperm whale oil and 30 barrels of right whale oil, having sent home 33 barrels of sperm whale oil.

Wednesday, November 25th

Raised whales on the 17th ultimo lowered and had a hard chase, and that's all the good it did us: all in good time no doubt, but such hard work and no pay does bring out our bad qualities. During the past ten days not much of consequence has occurred. We have had good weather and gammed for the second time with the ship *Brewster*, twice with the *Eugenia* and *E. Dunbar*, once with the *Eagle* and *Swallow*. Last Tuesday commenced *turning in* the main rigging, by 11 AM Wednesday six shrouds were turned, while engaged on the succeeding ones sperm whales were raised, about 10:30. Lowered quicker 'n Jarvis and our boat got two, one 60 bbl whale in the morning, and about 1 PM got fast to and killed a 40 bbl customer—the ocean to leeward seemed covered with spouts—an enormous shoal.

"Cruising Off Fort Dauphin" shows the *Clara Bell* in rough seas under jib, fore course, and reefed topsails, hunting sperm whales in the rough seas and strong wind off the southeast coast of Madagascar.

"Ultimo" means in the preceding month, but Weir seems to use it to mean the 17th past, or November 17th. The 225-ton ship *Brewster* of Mattapoisett, commanded by Captain Crary B. Waite, departed for the Indian Ocean whaling grounds on May 11, 1857, and returned on August 28, 1860, with 1,057 barrels of sperm whale oil, having sent home 83 barrels of sperm whale oil. The 336-ton bark *Eagle* of New Bedford, commanded by Captain John McNelly, departed for the Pacific whaling grounds on October 22, 1856, and returned with 930 barrels of sperm whale oil. The new 439-ton ship *Swallow* of New Bedford, commanded by Captain Herman N. Stewart, departed for the Indian Ocean whaling grounds on October 2, 1856, and returned on December 22, 1860, with 600 barrels of sperm whale oil and 890 barrels of right whale oil.

Saturday, November 28th

4:30 PM under short sail, abominable weather raining furiously, 7 PM cleared off finally, during the night had a splendid moon, this morning made all sail steering southwest, the weather is good.

Sunday, November 29th

Steering west by north, 3:30 PM raised a sperm whale, lowered at four, 4:45 Mr. Perry succeeded in fastening to a noble fellow, by this time the weather had become quite rough; Mr. Whale died slowly, though he had many mortal wounds.

Mammy Dinah shot two bomb lances at him, but one of which took effect. 7 PM had Mr. Whale safely tied by the tail with a two-and-a-half-inch cable. Commenced *cutting in* at daybreak, by noon, had the flukes and all blanket pieces on deck.

"Mammy Dinah" is a disparaging reference to Second Mate Walsh with his bomb lance gun.

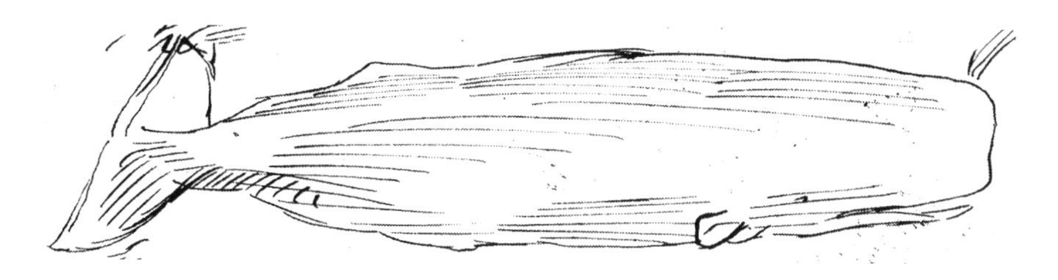

Monday, November 30th

3 PM got the junk on deck and while securing it (the weather being quite rough) the tail rope of the head chain parted and the case sunk. I must say I rejoiced secretly at our loss, for it saved us so much more hard work, and the loss of 10 bbls would be but a few pennies from my pocket. 6 PM the works were started, at midnight commenced to fill the casks, the blubber tries out finely. This morning Manuel Joseph a Gee got severely jammed by a loose junk cask filled with blubber sliding from its place upon him—weather improves.

Weir mentions some of the accidents that might occur in whaling, including the loss of a sperm whale case and its spermaceti, and an injury from a runaway cask on the slippery, oily deck during the trying-out process.

Tuesday, December 1st

Stowing between the run and booby hatches somewhat of a job, so many casks must be stowed full, on account of the shortness of the hose. 6:30 PM quit work though about half done. By daylight hard at it again, 11 AM finished, and such a dirty looking set of sailors are seldom seen. I am fortunate in having a watch below this afternoon, though we are liable to be called up at any moment for whales.

Because the canvas hose for filling oil casks in the hold was so short, they had to fill some on deck and maneuver them into the hold. The run is the after part of the main hold, where the vessel narrows toward the stern. The captain might have the more desirable provisions stored there, below the officers' quarters, to protect them from the crew. Later in the voyage, it would be an arduous task to maneuver oil casks into the run. A booby hatch is a deck hatch fitted with a ladderway and a sliding cover for access to quarters below. The hatch leading to the "steerage," just forward of the officers' quarters, where the boatsteerers, carpenter, cooper, cook, and steward lived, was generally called the booby hatch.

Wednesday, December 2nd

This afternoon gammed with the ship *Martha* and barks *Ann* and *Columbia*. Finished gamming about 8 PM, they leave for the Desolation cruising ground soon. This morning broke out water, rugged weather. The captain has prohibited us using water for washing.

The 299-ton bark *Ann* of Sag Harbor, New York, commanded by Captain Hamilton, departed for the Indian Ocean whaling grounds on December 7, 1855. The bark was condemned as unseaworthy at St. Helena on February 25, 1858, and sent home 380 barrels of sperm whale oil, 720 barrels of right whale oil, and 6,000 pounds of whalebone (baleen). The 285-ton bark *Columbia* of Sag Harbor, New York, commanded by Captain White, departed for the Pacific whaling grounds on April 25, 1856, and returned on May 31, 1858, with 143 barrels of sperm whale oil, 960 barrels of right whale oil, and 5,700 pounds of whalebone (baleen), having sent home 100 barrels of sperm whale oil. The Desolation grounds lay around the Kerguelen—Desolation—Islands in the south Indian Ocean about 2,000 miles southeast of Madagascar. Officially discovered in 1772, the Kerguelen Islands became a focus for sealing voyages and then for right-whaling and sea-elephanting voyages by the 1850s. The giant elephant seals were hunted on the islands, then their blubber was tried out for oil and their hides were saved for leather. With the supply of fresh water dwindling, Captain Robbins prohibited using it for anything but consumption.

Thursday, December 3rd

Nothing particular doing, during the night heading southeast, the weather is quite rugged, this forenoon steering west-northwest.

Friday, December 4th

Heading northeast strong easterly wind, real Fort Dauphin weather—that is bad weather. After shortening sail at sunset wore ship heading east-southeast 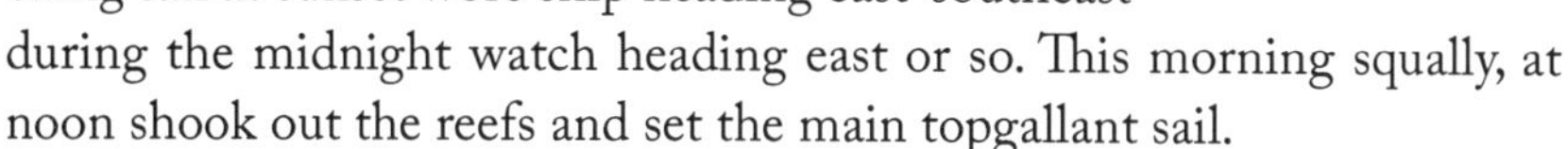 during the midnight watch heading east or so. This morning squally, at noon shook out the reefs and set the main topgallant sail.

Saturday, December 5th

Took in the main topgallant sail at 1 PM. 4 PM shortened sail to close reefed main topsail, spencer and staysail. Commenced raining very hard if not harder, hands employed in catching water during the night—managed to fill three casks.

Quite pleasant this morning, made all sail steering northwest, 9 AM steering north. Ten dollar reward for a 60 bbl whale, exert yourselves you Gee's.

To augment their dwindling fresh water supply, the crew probably used spare sails, stretching them horizontally to catch the rainfall and drain it into casks. It was common for a captain to offer a reward to the man who first sighted a whale. The incentive sharpened all eyes.

Sunday, December 6th

Steering north, excellent weather. *Wetting hold* as usual on Saturday afternoon. Chips has been repairing the gangway planking. Portuguese all aloft looking for ten dollars, Yankees schrimschawing, or whatever way these intolerable whalemen call it. At sunset luffed by the wind heading east by south. 12 PM wore ship heading south-southeast at 6 o'clock this morning squared the yards steering southwest by west, fine weather continues.

Wetting the hold was wetting down the oil casks stored on their sides in the hold to keep the staves from shrinking and the casks from leaking. "Chips" was the ship's carpenter, who was repairing the gangway planking where the whale head parts and large blanket pieces of blubber were dragged on board. Weir implies that the crew had separated itself into "Yankees" and "Portuguese." "Scrimschawing"—scrimshawing or scrimshandering—was a whaleman's hobby of carving implements or ornaments from ivory sperm whale teeth or incising sperm whale teeth or pieces of baleen with images, often as gifts for loved ones at home.

Monday, December 7th

Braced yards at 1 PM heading southeast. Of late we have taken the precaution to furl the foretopsail at nights in order to save the old sail as much as possible. 2 AM wore ship heading north-northwest. This morning watch employed in rattling down the larboard main rigging, pretty good weather.

While cruising about on the whaling grounds, it was not necessary to carry a full suit of sails. With the foretopsail showing signs of wear and weakness, it made sense to furl it at night, when approaching squalls could not be seen and a dash aloft was more dangerous. Rattling down the rigging means to replace the ratlines (footropes) secured to the shrouds as ladders for going aloft.

Tuesday, December 8th

Finished rattling down. Steering southwest, 12 PM wore ship heading west-northwest—this morning steering northwest, wind from east and south. At 10:30 AM great excitement—raised square heads and at the same time saw an enormous bone shark—as large as a 25bbl whale. We lowered and chased till noon, when we returned to the ship—crowded on every rag of canvas, gaining on the whales fast. I wish the sea was not quite so rugged.

"Our Bully Mate" shows Mate Barker in his attitude of supervision on deck.

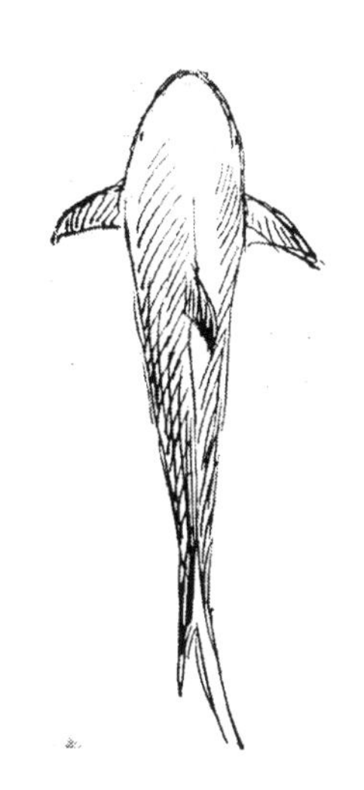

Wednesday, December 9th

Lowered a little before 1 PM, at 1:30 PM we succeeded in sailing upon a noble fellow, with studding sail and mainsail out our good boat fairly flew before the wind. When Barker the Pirate sang out "look out for him Wallace" I was up in a moment, let go the studdingsail tack, picked up my first iron and darted with might and main right in the center of Leviathans side. The second harpoon was buried in the bunch of his neck, in an instant he darted about three ships length ahead of our boat rolled over on his back, and worked his old jaw like the lever of a steam engine at the same time lashing the sea with fury. The starboard boat now ventured up, and with some difficulty, succeeded in getting an iron in to him, when Mr. Whale politely poked his jaw through the boat, making a hole as big as the head of a barrel.

"Oh! Dear oh! Dear I'm stove" says Mr. Perry, "never mind" says Barker, from our boat, "unbend your boats sail and plug it up."

We had to work warily about this customer, after considerable maneuvering Mr. Barker darted his lance, and in a moment the whale had our boat in his ponderous jaws and raised high out of water—crack, crash, crack—we were all tumbled pell mell into the water, and our boat left a total wreck, bitten to pieces—happily no one was injured. It was now time to look about for the ship. Mr. Perry could give us no help for he was already badly stoven, and could with difficulty keep afloat. There

Weir's largest journal illustration seems to be inspired by the taking of "Taber Tom," a massive sperm whale, though the presence of three boats in the drawing is more to provide artistic balance than to present an accurate depiction of the action. Weir would later use a similar composition for a large watercolor and for a *Harper's Weekly* illustration.

was the *Clara Bell* a mile off, dead to leeward, beating up under a stiff topgallant breeze. The men trimmed the sails with a will that day and our noble bark beat up gloriously, soon we saw the captain lower in the waist boat, but so soon as he came near the whale (who was scribing a circle about the whale all the time) Mr. Whale put towards him, this was a fix, however Mr. Perry pulled up with his stoven boat, and whilst he attracted the whale's notice the captain took us all safely on board and off we started for the ship, leaving Perry alone with the whale. It took but a few minutes to unlash a spare boat from the hurricane house and launch it—oars, tholepins and a spare lance were put into it, and off we started for the fight.

Mr. Barker took Mr. Welch and his bomb battery up on the starboard side of the whale while the captain took Mr. Perry up on his larboard quarter. Perry darted his lance with good effect and at the same moment Welch fired a bomb lance into him—which performances made his whaleship furious—the flap of his flukes upon the water sounded like artillery and his jaw came down like a trip hammer. The second lance Mr. Perry threw brought the blood from his spouthole. Mr. Welch fired three bomb lances into him, when Mr. Whale knocked off the head of his boat, and tossed our notable man with the bomb gun into the water.

His whaleship expired with the day, and by 6:30 PM we had him *tied by the tail.*

At sunrise we commenced *cutting in*, 10 AM body in. 11:45 junk safe on deck and now for dinner, with a desert of yarns about Taber Tom as the boys call our whale.

"Bomb Lance" shows the projectile fired from a bomb lance gun like that used by Second Matc Wclch. It appears similar to the bomb developed for the Brand shoulder gun, a 16-inch iron tube filled with gun powder, with a lance-like head and rubber fins to stabilize its flight. A shotgun shell propelled the bomb lance out of the gun and ignited a fuse, causing the bomb lance to explode in the whale, producing lethal damage.

Weir's whaleboat probably set a four-sided spritsail, and it is not clear what form of studding sail would serve as an extension of that sail. Captain Robbins described this incident in his book, *The Gam*. From a distance, the captain saw Weir in action: "There is frantic excitement in the mate's boat off to leeward—'Stand up and give it to him! Quick, quick, *quick*!' See!—a figure erect in the boat's bow—a long shaft wielded in both hands high over the man's head—a momentary poise—a swift springing motion—a sudden recoil—the harpoon is hurtling through the air—the slender line singing after it—the weapon sunk fast in something, and that something sinking rapidly into the depths, dragging the line through the chocks so fast the druggs (Robbins seems to refer to a "drug" or "drogue," a bucket or plank attached to the line to slow the whale) could do nothing to steady it—fifty fathoms—a hundred—two hundred! The mate and the harpooner have changed places. The men dodged the flying line. Now followed a fresh period of suspense—anxious but brief. . . Suddenly and all unexpected, the whale came up again like a submarine boat. He bumped his back against the blades of the first mate's oars. His shiny black hump stood fairly a foot out of water. The men could feel the damp heat of his spout. We could hear the sound of it. . . . He lifted his huge square nose ten feet into the air, and dropping his long under-jaw, deliberately calculated his distance. Then with a hideous swing of his whole appalling mass, he veered round and took that whale-boat into his mouth. His ivory teeth smashed through the cedar clinkerwork. The boat went to pieces like an egg-shell. The mate's crew flung themselves into the water, and escaped by the skin of their teeth." Weir confirms that the whaleboats were fitted with wooden thole pins rather than iron oarlocks. With the captain and mates in the replacement boats, Weir then went as an oarsman while the captain and mates succeed in lancing the whale. "Deafening indeed were the cheers from the ship's deck when we had won that desperate fight," noted Robbins, and "when they took the falls to the windlass and manned the bars it was a joy to hear them sing." Robbins concluded: "What a whale that was! He was the biggest fellow I ever fell in with. He measured sixty-four feet overall, and he had a sixteen-foot jaw. His flukes stretched sixteen feet from tip to tip. He made a hundred and thirty barrels of oil" (Robbins, *The Gam*, 172–73, 177–79, 180). The significance of the name Taber Tom is not known.

Thursday, December 10th

3 PM case safe on deck, and a couple of men buried to their necks in the centre of it, bailing out the spermaceti as though a human life depended upon their exertions—wouldn't this be a rich scene for the dear ones at home to see, a couple of men buried in a whale's head—a delightful? situation surely.

Started the works between 4 and 5 PM trying out the head first. 6 PM sail in sight chasing whales, too late in the evening for us to lower, though the whales could be seen plainly from deck. Hard at work with the fires all night. This morning the bark *Montgomery* close to, chasing whales. 8 AM her boats fast, we lowered and chased for a couple of hours, without success. All hands on deck, from sunrise to sunset, five and a half hours rest out of the 24, pleasant and no mistake, the old skipper takes it comfortably and gets well paid for it, but the officers suffer with the sailors—good.

"Sports of Whalemen" shows another instance of a sperm whale crushing a boat just after the boatsteerer has darted his irons.

Friday, December 11th

The blubber tries out well, whales going to windward fast. 8 PM spoke the bark *Montgomery*—350 bbls 3 years out, unfortunately. This morning broke out sail and bread casks to put oil in; this looks like driving the business—all our casks for oil are filled, all hands on deck, getting our new boat in readiness for placing on the cranes.

With all the dedicated oil casks full, the casks for storing spare sails and "bread" or hardtack crackers were emptied to fill with oil. Mate Barker was given one of the spare boats carried on the overhead skids.

Saturday, December 12th

Shifted the bow boat from the larboard boats cranes, and set the new larboard boat at rest, and now I have got business enough for a week to get her in good whaling trim—to *set up* five irons, three lances and all sorts of whaling tackle to be stowed in her—strangly [?] 30 bbls *turned up*, since Thursday evening.

Apparently the bow boat had been hung in place of the lost larboard boat after taking Taber Tom. Now the crew got the new larboard boat hung from the davits, resting on the cranes. As boatsteerer, Weir was in charge of setting up the new boat with line tubs, whaling irons, and the other components for active whaling.

Sunday, December 13th

Finished trying out this afternoon—75 bbls. We have good weather but whether it will continue is doubtful. One sail in sight this morning, supposed to be the *Montgomery*—somewhat rough.

"A Case (Upper Part of a Whale Head Containing Spermaceti)" shows a sperm whale case lashed on deck as the boatsteerers bail the liquid spermaceti from the cavity within. Weir shows the whale's S-shaped blowhole atop the case at left. One man spills a tub of spermaceti as he loses his footing on the slippery deck.

Monday, December 14th

Sail in sight proved to be the bark *Afton*, ran off, spoke and *gammed* with her—18 and a half months out, 900 bbls sperm, on her passage around the Cape—this vessel took our oil home from St. Helena, 1855. This morning hard at work stowing down the oil, under the fore peak.

Tuesday, December 15th

Finished stowing the oil by 4:30 PM and washed decks, wore ship at sunset heading southeast, wore round 12 PM heading north. Kept out jib, spanker and mainsail all night, pretty good weather this morning.

Wednesday, December 16th

At 2 PM bent the new main topsail, after which it rained a little. Watch employed in breaking out apples, rice and pickles—only think of such luxuries so far from home. At sunset wore ship heading southeast by east. Wore 'round again at midnight, heading north-northwest, it was dead calm and took us *near* three quarters of an hour to bout ship. This morning engaged in reeving new braces, buntlines, clewlines, spilling lines, slab lines etc. Beautiful weather.

"To bout ship" means to bring it about from one tack to the other (here by wearing, or turning the stern through the wind). Weir mentions the different lines attached to each square sail: braces to swing the yard, buntlines to raise the foot of the sail, clewlines to draw up the clews in furling sail, spilling lines to raise the entire body of the sail, and slab lines to raise the foot of the fore and main course as needed.

Thursday, December 17th

Repaired our boats steering oar; good weather; heading northwest by north a half north. At sunset wore ship, heading southeast, weather begins to look bad. 2 AM wore ship, heading northeast a half east—variable wind. Made all sail this morning steering northeast and northeast by east. Repairing the old topsail.

Friday, December 18th

Steering northeast, fitted up the lantern keg for our boat, still repairing the topsail.

During the night the weather looked quite threatening. 3:30 AM commenced raining, wind unsteady, head about north-northeast, set full topsails at daylight. 6:30 AM wore ship heading southeast bad looking weather. Seized new reef tackle blocks on the fore-top-sail yard; enjoyed rainy and squally weather towards noon.

Among the equipment in a whaleboat was a small, conical wooden keg that contained a lantern as well as some pieces of hardtack for emergency use if the whaleboat was separated from the vessel. Reef tackles were used to gather in the edges of square sails in the process of reefing or shortening sail. The lines ran through blocks at the mast and then down to the deck.

Saturday, December 19th

Villainously rainy, standing single boats crew watches, making worm pegs and schrimschawing in general, heading anywhere between south and southeast. Towards evening the weather improved a little. Kept jib and spanker set all night. The steward caught an enormous dolphin, the grains were darted but the line slipped from *somebody's* butter fingers, letting grains dolphin and all go to Davy Jones. Made all sail this morning broke out three casks of sails; stowed the sails in the run; and our indomitable cooper is preparing the casks for oil, we also broke out water and filled the scuttle butt. The weather is still somewhat rainy, but not enough to let one boats crew go below, or to keep the sailors idle. Steering southwest, somewhat squally, at noon luffed on the starboard tack heading northeast.

Dating at least to the early 1700s, the concept of a monarch of the depths named Davy Jones has been common among sailors. Sometimes described as a devilish creature, sometimes associated with the Biblical prophet Jonah, who was swallowed by a "whale," the name Davy Jones may be a combination of St. David, patron of Welsh sailors, and the prophet Jonah. Drowned sailors resided in Davy Jones's Locker. Like most provisions and stores, spare sails were brought to sea packed in casks. Lashed on deck, the scuttlebutt was a wooden cask holding about 126 gallons, with an opening in the top for access. This was considered a day's allowance of water for the crew, and each day the scuttlebutt would be refilled from the water casks in the hold. Especially in the navy, sailors might gather at the scuttlebutt for water and to share gossip and rumors, leading to the use of the word scuttlebutt for gossip.

Sunday, December 20th

Washing ship, considerable rain, and unpleasant withal. At sunset wore ship heading south-southeast and south by east. 2 AM clear weather, wore ship heading north-northeast. At sunrise made all sail, steering southwest, 10:30 AM altered the course west by north.

Monday, December 21st

Took in the light sails, stiff breeze from the eastward. Steering west-northwest, at sunset luffed to the wind heading southeast by east. 3 AM wore round heading north by east. 6:30 AM steering north repairing the main lifts, grinding and cleaning the spades, to be in readiness for cutting in. At noon land in sight (Madagascar) tacked ship heading south-southeast —Cape St. Mary's astern.

Lifts are the diagonal fixed lines from the yardarms to the mast that keep the yards level. Whaleships carried a hand-cranked grindstone for sharpening implements such as the cutting spades used for severing the head parts and blubber from whales. Cape St. Mary, now Cape Sainte Marie, is the southernmost point of Madagascar.

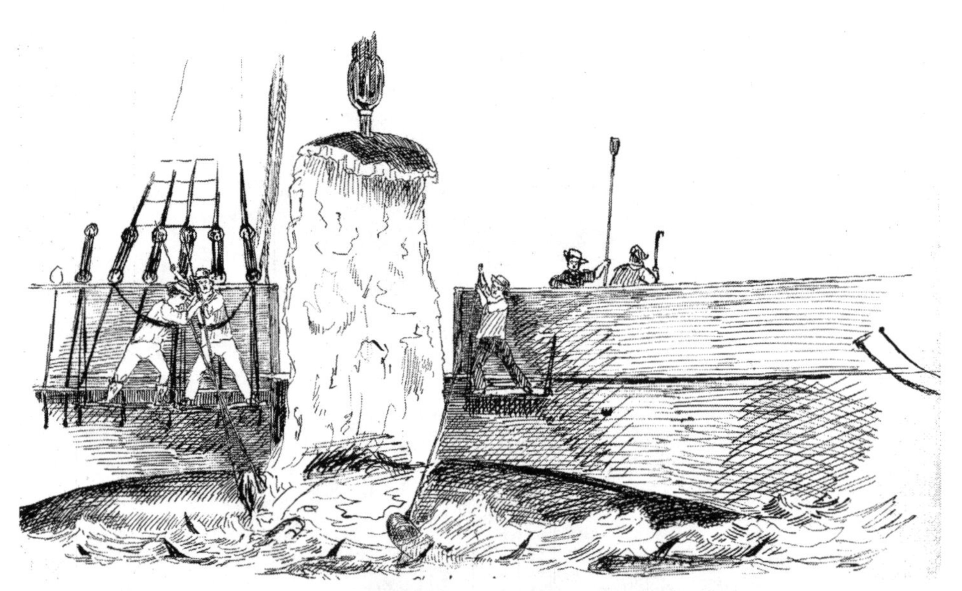

"Cutting In" shows the mates standing on cutting stages hung outside the rail, using their long-handled cutting spades to cut free the blanket piece of blubber in a spiral from head to tail. The lower block of the cutting tackle is attached to the free end of the blanket piece. A sperm whale's case and junk were cut free in the same way. Notice the sharks gorging on the carcass. This sketch also shows the arrangement of deadeyes and chain-plates at the main chains.

Tuesday, December 22nd

Heading south-southeast, seized new reef tackle block on the main topsail yard arms; in the evening distributed one last keg of tobacco, abominable stuff 30 cents per pound. At sunset shortened sail, heading south by east, 2 AM wore ship, heading northeast; overhauled and repaired the cutting falls and blocks, sent down the old fore topsail and bent a new one.

Wednesday, December 23rd

Heading northeast by east, pretty good weather. 6:30 AM Mr. Perry raised sperm whales, about four miles off, 9 AM the waist boat got fast to a 40 bbl bull, went upon him as he was breaching: a few of the whales brought to for a moment and Mr. Welch fired a bomb lance into a loose whale which we chased till noon with no success.

This bull handled his flukes admirably, our two boats *held off* and watched the contest and Mr. Welch acquitted himself well—he avoided his flukes with such good skill, and at the same time gave the whale such severe lances, that he soon put about in his flurry and then we had to scatter, and when Mr. Whale lay *fin up* we saluted Mr. Welch with three hearty cheers—the success of that battle will be uppermost in his mind for many a day.

Thursday, December 24th

1:15 PM commenced *cutting* in, 4 PM all valuable parts of the whale on deck and the carcass floating away, 5:30 PM started the works, trying out the case, weather not very favorable, 7:30 AM raised sperm whales off the weather bow, made sail and lowered immediately, at 9 o'clock, Mr. Barker put me on a noble fellow. We were right over the whale's flukes and I saw he was going down. I did not forget that the captain had said "two whales more boys and we'll go home." I darted the iron right below his hump, I heard a crash behind me, and was covered with a deluge of water, but I saw nothing but the whale. I thought his carcass would bring us so much nearer home—the second iron was darted home and I turned around—what a sight—all the crew amidships piled up like so many potatoes, the boat half full of water, and the line running out like lightening, smoking

"Going Over the Whale's Flukes" shows the activity on December 24, 1857, with Weir darting his second iron as the whale lifts the boat on its flukes. Although the boat was damaged and full of water, by the time the whale resurfaced the crew was ready and Mate Barker lanced it, with the help of Mate Perry and his boat.

like a house afire—no time was to be lost; the men tore off their frocks and shirts and stuffed them in the hole; at the same time baling boat with might and main—the whale meantime came up and Mr. Perry fastened to him and *gave him a lance*—we had now got our boat pretty clear of water, though two men had to keep constantly bailing, we hauled up to the whale and the mate gave one of his true darts, which made the blood run in a stream from the whales spout hole. We now pulled outside of his circle and waited for him to die which he did in about half an hour, his flurry was terrific but glorious to behold.

Friday, December 25th

1 PM Mr. Whale secure alongside; saw more whales close by, quarter boats chased, no use. So we came alongside a few minutes before 2 o'clock, and after eating a hasty dinner set up the falls and were fairly at work cutting in by 3 PM. 7 PM had head and all on deck. The blubber looks *remarkably fat*, broke out casks—at 8:30 PM started the fires again for we had to cool the works while cutting. Horrible weather, rainy and squally—the works are retarded much. All hands on deck today, peeling

jaw, cutting junk, bailing case etc. etc. And Christmas day too—this day of all others to be in such a pickle. I'm glad father can't see me and yet I'm sorry too, but I hope they are all happy at home now—they can't have the slightest idea, how pleasantly I am enjoying this blessed day.

Saturday, December 26th

Got the decks pretty well cleared—we are driving the fires at a fearful rate, and those poor fellows whose work in the waist cutting back bones, fins and junk can smoke without using tobacco or pipes—though much to the injury of their lungs and eyes. Wretched weather during the night—rainy and squally, spread an old topsail for an awning and three minutes after, it was blown away. Oh! What a jolly life.

Weir alludes to the heavy black smoke produced by burning blubber scraps in the tryworks. To keep the rain out of the boiling oil, they tried to rig an awning over the tryworks but it blew away.

Sunday, December 27th

Sail in sight this morning; the weather is dull, broke out several water casks, three casks of sails, a cask of slops and one of towline. Stowed sails, towline and slops in the run, broke out water—the blubber tries out fast and well. 7 PM commenced trying out head, 7 AM commenced to put the *fat-lean* and *slumgullian* in the pots: raining an old English rain all the morning, our clothes soak it in in spite of the coating of oil they have got. Upwards of 115 bbls on deck, this looks something like whaling—what a time we'll have stowing it away, but it has got to be done.

With so much oil tried out, the crew needed to appropriate more supply casks, containing sails, clothing, and equipment to be sold to the whalemen, and whaleline, and even water casks from the hold to contain it. Slumgullian means the scraps and oily, wet material left when the blubber was cut up.

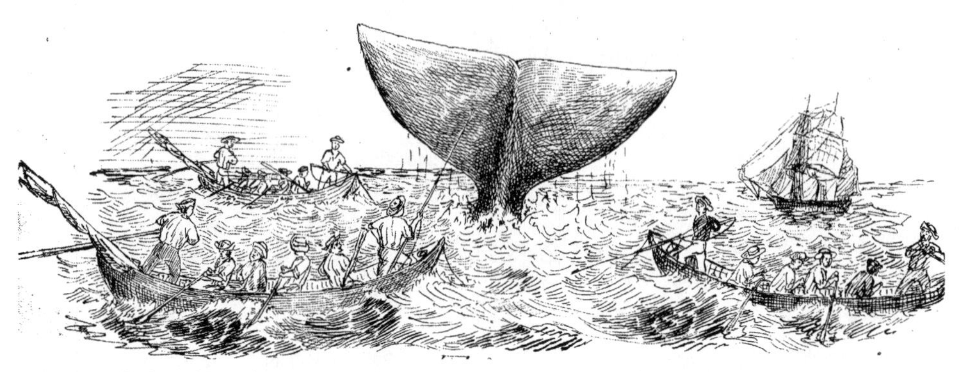

"Too Late," three boats' crews watch as a sperm whale turns flukes and sounds.

Monday, December 28th

Heading north-northeast under double reefed topsails, plenty of rain with little wind, decks scrubbed, oil cooling preparatory to stowing it down, and all begin to think of home once more, though the skipper still gives the cry "two whales more." 6 PM land in sight, (Madagascar) wore ship heading southeast quite cloudy and threatening during the night. Made all sail this morning, hands engaged in coopering the oil. 9 AM tacked ship, steering north-northwest, sail in sight supposed to be a merchantman, her course is about southwest. I wonder when we'll steer such a happy course.

Tuesday, December 29th

Heading north making preparations for stowing the oil early tomorrow morning. 6:45 PM shortened sail and wore ship heading southeast, Fort Dauphin weather. This morning hard at work stowing the oil under the fore peak and in the forward part of the fore hold. The weather is growing pleasant.

Wednesday, December 30th

Finished stowing the oil by 5:30 PM after a hard day's job—kept the course, jib and spanker set during the night. Cape Saint Mary Madagascar in sight last evening, during the night heading southeast by east. This morning after making all sail, scrubbed decks. All hands including the captain assisting the cooper set up some old shooks that seem fit only to burn, however with pretty hard scratching we can muster enough casks to contain 200 bbls. 8 AM squared the yards steering south, wind from north.

Shooks were bundles of staves, with heading and hoops, from which a barrel or cask was reconstructed by the vessel's cooper. Bringing empty casks to sea in the form of shooks saved space on board. These last casks seem to have been made up from leftover staves in poor condition.

Thursday, December 31st

Steering south with moderate northerly breeze—5 PM steering south-southeast, looks squally. After taking in sail luffed by the wind on the larboard tack, heading northwest by west, wind variable from southward and westward. This morning steering northeast, coopering shooks and looking for 150 bbls more to take us home. The last day of the declining year.

1858

Friday, January 1st

Heading about southeast during the night. Made all sail this morning, a splendid day, the dawning of the New Year, and here we are hard at work coopering on the decks of *Clara*. Precious few calls I make this day.

Steering northwest with a pleasant breeze—9 AM the martingale guys gave way, by 10:15 they were repaired with preventer guys added. 11:15 AM raised sperm whales and chased, with all sail set till near noon when we lowered away the boats, the whales had gone down, and we were pulling industriously towards a spot where we thought they would rise—when to our surprise a pod of them rose directly under our boat. I grasped the harpoon to heave into the side of a noble boy fellow when I heard the mate crying out in a rather loud whisper; "Are you mad man, you'll kill us all." I looked around and here we were almost resting on the flukes of one and the head of another—the slightest prick to any one of a half dozen of whales about us would certainly have been our ruin. I momentarily expected the boat would have been dashed to atoms—there I stood harpoon in hand, and the crew with oars suspended in the air, like so many statues; did not dare more a muscle, when the whales became aware of some danger, were gallied, and all sunk like so many tons of lead. We now breathed freely.

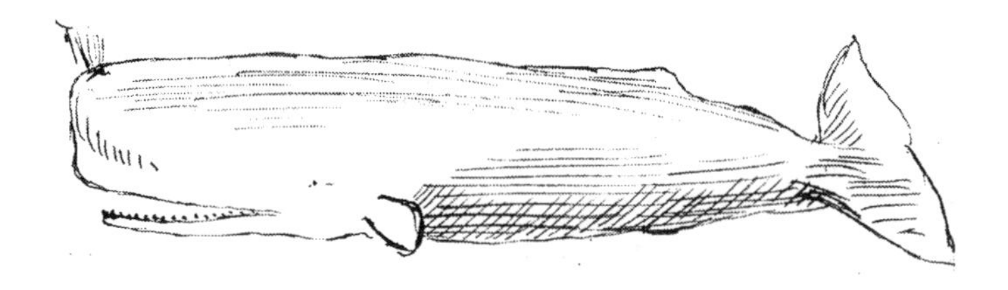

Saturday, January 2nd

We took the whales on the next rising—the starboard boat got fast shortly after meridian, we were not long in following—the whale ran a great deal, the Old Man fearing to lose the whale sent Mr. Welch from the ship with the bomb gun, but before he reached us Mr. Barker had given the monster a lance which made the thick blood flow from his spout-hole—take your bomb battery home again most valorous Mr. Welch. Turned the old fellow up in a short time, by 3 PM had him safe alongside and all hands below for a few moments to dinner.

After dining commenced *cutting in* with a will, the blubber hook tore out—came near knocking a half dozen of us over—blubber guys kept constantly parting, everything seemed to work contrary, and all from the skipper down to cook, seem to be in bad humor. Sail in sight, proved to be the *Montgomery* she bore down and spoke us in the midst of *cutting*; at 7:30 PM left off work in disgust, after all our exertions only got jowl and jaws on deck. 5 o'clock AM commenced cutting again, by 8 AM flukes in the blubber room and all hands at breakfast. Sail in sight off weather quarter bearing toward us, turned out to be our old friend the *John Dawson*, their mate Mr. Albert came on board and the Old Man left with my crew for a *gam*. By 9:30 AM case and junk on deck, by noon all things ready for starting the works, dull and rainy weather. Mr. Albert and his crew have been assisting us all the morning while our boats crew are taking things easy in the other ship.

First Mate William C. Albert of the *John Dawson*, a native of Tiverton, Rhode Island, had gone to sea at age 18 on the bark *President* in 1844. In 1850 he was second mate of the bark *George Howland*, followed by three voyages on the *John Dawson* as first mate. As Weir suggests, during a gam, the captains gathered on one vessel and the first mates and their boat's crews gathered on the other vessel.

Tuesday, January 3rd

Started the fires at 1:45 PM, boiling head by 12:45 midnight, 8 bbls of pure spermaceti turned up. 7 o'clock this morning finished the head and

"Heave Pall" shows the crew at the windlass, pumping up and down on the long levers of the pump break mechanism that turns the barrel of the windlass. The pall (pawl) is a heavy wooden fitting that drops into notches in the windlass barrel as the crew pumps, keeping it from turning backward. The angle of the lines at right suggest that they are hoisting the cutting tackle to bring whale parts on board. At center is the forecastle companionway.

started on the body. Favorable weather—under double reefed topsails, heading about northwest. *J Dawson* still in sight—they'll probably stick by us now in hopes of getting some luck.

Today the dear folks at home will be going to church and I poor devil must be driving away at these fires—what a barbarian I'll be to show myself among ladies again. I wonder if anyone at home can entertain half an idea of the luxurious and easy life of a Whaleman!

Monday, January 4th

John Dawson not above a mile to the windward of us. Getting on pretty fairly with the works. 11 AM commenced raining like Sancho, had to *cool down*—it came down in torrents during the midnight watch, and we left the pots in an awful mess. What they contained was a fair sample of the most glutinous glue—the wind shifted continually. At 4 AM started the fires again, and such a pickle I never want to poke up again. Made all sail this morning. Raised fin backs breaching—12 noon soup for the million in the pot.

Trying out blubber extracted the oil from the connective tissue and also evaporated much of the water that emerged from the flesh with the oil. If it rained during trying out, the tryworks was often sheltered with canvas to keep the rainwater out of the trypots. Apparently, they let the fires die and the oil and rainwater solidified into a gummy mess in the pots. When sperm whale oil was finally processed at an oilworks ashore, it would be boiled for about 10 hours to evaporate the last of the water mixed with the oil.

Tuesday, January 5th

Commenced this nautical day by tacking ship, cleared up the decks, wet hold, extracted the teeth from the whale's jaw. After supper baled the pots, cleaned the cooler etc. etc. Heading north-northeast—tacked ship at midnight heading east, wind from north and east. This morning coopering the oil, cleared away in the fore hold for stowing—plenty of fin back whales about.

A sperm whale has about 40 teeth in its lower jaw, which fit into sockets in the upper jaw. It is believed that sperm whales use their teeth to seize their principal food, giant squid. Often the jaw would be saved during cutting in, and later the teeth were extracted for use in the whaleman's art of scrimshaw. From captain to greenhand, men might engrave the teeth with designs or carve them into ivory implements or decorative items. The wide, flat panbone of the jaw might also be used for scrimshaw.

Wednesday, January 6th

Finished stowing the fore hold, coopering and jobbing in general. Variable winds during the night, 2 AM heading west-southwest, wind from southward. This morning hard at work stowing the oil on the top tier starboard side between the fore and main hatches, steering west.

"Extracting Ivory" shows whalemen stripping the teeth from a sperm whale's jaw, to be used for scrimshaw.

Casks were stowed in the hold on their sides in fore-and-aft tiers, with the bungs up, with wooden blocking to keep them from shifting. Since a gallon of whale oil weighs just over seven pounds, a full 31½-gallon cask would weigh about 250 pounds, making for a hard job in lowering and skidding them into place in the hold.

Thursday, January 7th

Stowed four casks in the after hold. 4:30 PM decks cleared. Running off for a sail, 5 PM gammed with the bark *John Dawson* till 7:30 PM, during the night heading east-southeast. Sail in sight this morning. Making and taking in sail eternally during the day—that seems to be the only employment the skipper can turn the hands at. Heading south by west. Wind somewhat variable.

Friday, January 8th

Heading south-southwest with very unpleasant weather, ship leaking badly. 10 PM heading south-southeast and southeast. 2 AM wore ship heading north by east, miserable weather, lots of fin backs about. The Old Man thinks that whenever these whales make their appearance sperm whales will be scarce—I don't care if it is so. 6 AM set full topsails, though

"Trying Out" shows the tryworks, just aft of the foremast and fore hatch. Weir often worked here, so he may depict himself forking "bible leaves" of blubber into a trypot. The man at his left stirs the fire, perhaps adding more "cracklings," or pieces of rendered blubber, to burn in place of wood. The heavy black smoke blows aft, filling the working area in the vessel's waist. A water-filled "goose pen" under the brick tryworks protects the deck from catching fire. Next to the tryworks is a spare trypot, stowed upside down, and a square copper cooling tank in which rendered oil, dipped from the trypots, was cooled. The man at left appears to transfer oil from the cooling tank to a large cask. The whaleman in the foreground seems to be scooping up puddled oil. The wheel in the background may be the "sailors' organ" for spinning spun yarn, a hated task.

it is squally, and now the old skipper says he is ready for the last whale on this abominable ground. At noon tacked ship heading southeast.

The finback or fin whale is a large rorqual whale, a filter-feeder with pleated flesh under its jaw for expansion as it takes in water to force through its filtering baleen. Finbacks were rarely hunted by American whalemen as they swam fast, usually sank when killed, and did not produce much oil or baleen.

Saturday, January 9th

Chips our carpenter is making a new topmast studding sail boom from a spare to'gallant spar. 6:45 PM tacked ship heading north by east. 2 AM wore ship heading southeast. 6 AM tacked ship heading north, 10 AM wore ship heading southeast, at noon tacked heading north—pumping ship quite often, serving the main top-mast backstays, at noon set full topsails.

Sunday, January 10th

Heading north, wet hold as usual on Saturday afternoon. At sunset wore ship heading southeast and south-southeast. 6 AM wore 'round heading north by west, pretty fair weather. I wish I could be at the little church today.

Monday, January 11th

1 PM tacked ship heading east-southeast at sunset, double reefed the topsails, but kept out courses, jib and spanker. 10:30 PM tacked ship heading north a half east. Piled on all sail this morning, still at the pumps, making another top-mast studding sail boom from a spare spar. 6:30 AM sail in sight, watch employed sailorising—turning in the main top mast back stays. 9:30 AM gammed with the ship *Brewster* of Mattapoisett, Captain Waite, 150 sperm, fine looking craft but too small to be cruising so far from home.

"Junk" shows the portion of the sperm whale's head, below the case and above the jaw, that was also hoisted on board. Its spongy flesh was saturated with spermaceti and apparently played a role in the sperm whale's echolocation. Using cutting spades, three men section it for the tryworks.

Tuesday, January 12th

Still gamming with the *Brewster*, they have had quite a tragedy on board—one of their sailors a Kanaka attempted to kill all the Portuguese on the ship—he succeeded in stabbing five of them, two quite seriously. 8 PM finished gamming. Heading northeast, 2 AM wore ship heading south-southeast, 7 AM tacked ship heading northeast —*turning in setting up* and serving the topmast backstays. Two sails in sight, moderately good weather.

A "Kanaka" was a native of the Pacific islands, a term derived from the Hawaiian word for "people." Like young Azorean and Cabo Verdean men, young Polynesian men also went whaling on New England vessels in large numbers. This incident reinforces Weir's observation that there was racism or ethnic conflict among some whaling crews. By "turning and serving" Weir may mean weatherproofing the shrouds. A stay would have lengths of spun yarn or small seizing line laid into the grooves between the strands of the stay. This was called worming.

Then a strip of canvas was wrapped in a spiral around the stay in the direction of the spiral of the strands. This was called parcelling. Then the sailors would "turn and serve the other way," making a wrapping of thin, strong hemp marline around the length of the shroud in the opposite direction, using a special serving mallet for a tight wrap. This was called serving. When tarred, this covering protected the stout hemp shrouds from damage and rot.

Wednesday, January 13th

3 PM gammed again with the ship *Brewster*, 7:30 PM finished gamming and wore ship heading northeast. 2 AM wore 'round heading south-southeast, this morning, the ship *Brewster* in sight. We have agreed to exchange officers and one or two men, Mr. Welch packing up.

The *Brewster* had been out for less than a year and now needed several sound men to continue the voyage. The *Clara Bell* was nearly full and preparing to head home, so officers and crew members who wished to remain at sea were available. Captain Robbins was willing to give up his bomb-lance-wielding second mate, so David Welch transferred to the *Brewster*.

Sunday, January 14th

Watch employed in the highly edifying and very lucrative employment of splicing yarns for spun yarn. 3 PM *Brewster hauled aback* about three miles off our lee bow, and soon sent a boat alongside of us. Mr. Clark their mate, a perfect gentleman, came with his *furniture*. Auguste and Allan, two of our Creoles, went aboard the ship with their donkeys. At 7 PM Mr. Welch our second mate left us, bag and baggage. There were no tears shed and so we finished the gam. During the night had bad weather, rainy and squally, heading south-southeast and southeast. 7:30 AM tacked ship heading about north at noon—two sails in sight.

When hemp rigging wore out, it would be saved, then picked apart, and the fibers would be spun into thin strands of spun yarn, which could be used for many purposes on board. Hauled aback means that the square sails are braced to take the wind on the forward side, to stop the vessel. To hold its position, a vessel might brace the mainmast square sails to be partially filled and brace the foremast square sails the other way, with the wind on their forward side, to keep the vessel stationary. Mate Clark was 31-year-old Cyrus E. Clark Jr. of Fairhaven, Massachusetts. He had gone to sea as a 19-year-old greenhand on the *Canton II* in 1845, then served as a boatsteerer on the *William & Eliza* in 1848. He spent 18 months as master of the *Tropic Bird* in the South Atlantic before signing on the *Brewster* as first mate. "Creoles" Auguste and Allen may have joined the crew in the Azores as they are not on the initial crew list. By "donkeys," Weir means their straw-filled mattresses, sometimes called a "donkey's breakfast."

Friday, January 15th

Steering west-southwest. 3 PM gammed with the bark *Nautilus* Captain Hardwicke, of Nantucket and her crew are a good specimen of the island. She is six months out, 80 bbls sperm. 6 PM, while in the midst of a jolly gam, the bark *Montgomery* bore towards us from the windward making some unintelligible maneuvers. We piled on the canvas, and chased for an hour when we found Captain Chapman was fooling for a gam, he met but a cold reception when he boarded us. 7:30 PM finished gamming, Creole George was discharged and paid, he left us for the *Nautilus*, and young Adams, the son of the owner of the *Nautilus*, came on board of us to go home—it shows the time cannot be far distant. During the night heading northeast by north, 6 AM wore ship heading south-southwest, at 8 AM steering southwest—one sail in sight. Working on the studding sail gear, affairs at present look suspicious. Everyone is speaking of *home, home, home*—shall we ever see home?

The 220-ton bark *Nautilus* of Nantucket, owned by Zenas L. Adams and commanded by Captain Edwin M. Hardwick, departed for the Indian Ocean whaling grounds on June 22, 1857. The vessel sent home 80 barrels of sperm whale oil, and sometime after young Adams joined the *Clara Bell* and "Creole George"—again, probably a man who had come aboard in the Azores—joined the *Nautilus*, that bark was wrecked at Fort-Dauphin (now Taolagnaro) on the southeast coast of Madagascar.

Saturday, January 16th

Steering west-southwest, making spun yarn, grinding most sublimely awful notes from the sailors' organ—6 PM wore ship heading northeast, two sails in sight at sunset. This morning heading northeast by north, one sail in sight, unknown. Strong wind from the east and south manufacturing homeward bound gear—repairing blocks etc.—double reefed topsails.

The "sailors' organ" was the hand-cranked machine with spinning hooks that was used to spin old rope fibers into strands called spun yarn. "Organ" refers to its vague resemblance to a musical hand organ.

Sunday, January 17th

Heading northeast by north, seizing the main-top-mast back stays—two sails in sight. 6:30 PM wore ship heading south-southeast, by morning heading south by east. At noon wore ship heading north by east, fresh easterly wind blowing—thoughts running on home.

Monday, January 18th

A very heavy swell on, there must have been a severe storm to the eastward of us. At sunset furled the fore-top-sail and wore ship heading southwest, still obliged to pump ship quite often, 6 AM wore ship heading northeast by east. This forenoon bent a new foresail and jib. At 8:30 AM while bending the jib gammed with the *Jos. Maxwell* 26 months out, 1,250 sperm. They would do well to go home.

This sketch of the *Clara Bell*'s stern shows the "house" aft by the wheel, and the larboard whaleboat hanging from the davits and supported by the cranes. Like most vessels, the *Clara Bell*'s name and home port are listed on the stern. Many whaleships had a set of small stern windows opening on the captain's day cabin, but according to Weir's sketch, the *Clara Bell* did not.

Tuesday, January 19th

Still gamming, fresh breeze from the southward, heading southwest a half west. At 1 PM raised sperm whales, blowing pretty smart, rough sea on. At 2 PM lowered away the boats, Mr. Perry in the starboard boat, Captain Jenny in the larboard and Captain Chapman in the waist boat and I to steer him. Mr. Perry got on a whale and Frank threw two irons and missed—so much for a Gee, a most valorous Gee.

Finished gamming at about sunset and wore ship heading southeast a half east. At 2 AM wore ship heading west-southwest, sent down the foretopgallant yard and repaired fit for studding gear.

7 AM steering a course west, 9:30 AM sent up the topgallant yard, 11 AM wore ship heading southeast by south. Rove new fore braces and topgallant sheets. We have really begun to get the ship ready to go home—hurrah.

Some captains always lowered, commanding the starboard boat. Others preferred to remain on board and manage the whaling process from the vessel. Captain Robbins was eager to lower for whales. In his book, *The Gam*, he recalled waiting for a whale to surface: "There is something delicious in this exciting uncertainty. It makes your blood tingle. It makes your nerves thrill. It makes you feel yourself ready to face the whole world of perils and proudly conquer them all. You stand in the stern-sheets, leaning on the steering oar, and as you look into the faces of those five stalwart men on the thwarts before you, you tell yourself they are fine heroes, every man Jack of them. Yes, heroes! Soldiers face no greater perils. . . . So, as I was saying, you stand and wait, a-tingle with enthusiasm. You are in your glory now. You would not for the whole world be any other thing but a whaleman. . . To its toils and its perils you willingly devote your youth and best manhood. You will be proud, in long years to come, to recount the history of your daring sea-battles" (Robbins, *The Gam*, 169–70).

Wednesday, January 20th

Sent down the main topgallant yard and repaired the gear. 4 PM sent it aloft again. Two sails in sight.

At sunset shortened sail, furled the fo'topsail—the weather looks very threatening, during the night heading southeast by south, this morning heading southeast, fixing the homeward bound and studding sail gear etc. etc. 10 AM tacked ship heading west by south.

Thursday, January 21st

Fixing *homeward bound blocks*. 3:30 PM squared the yards steering northeast, fresh breeze from the southward. 5 PM luffed on the starboard tack, heading southeast by south. This morning we have fine weather, squared yards steering north-northeast looking for the *John Dawson*, bent and rove some new rigging, overhauling the royal yards, one sail in sight thought to be the *Montgomery*. 10:30 AM steering north, all hands talking about home—the prospect is that before many months we'll be there.

Dog Watch" depicts one of the watches at leisure during the 4:00–6:00 or 6:00–8:00 PM dogwatch period. With freedom of the foredeck, they gather around the windlass to smoke, sing, and tell yarns, or stories.

Friday, January 22nd

Steering north, overhauling the royal yards, sail still in sight—at sunset luffed to the wind heading east-northeast—squared the yards this morning steering southwest and west-southwest. We're soon for Madagascar.

Saturday, January 23rd

Steering west, finished the royal yards—sail in sight at sunset. Luffed heading southeast, wind from east. 5:30 AM wore ship steering north-northwest broke out molasses etc. etc.

8:30 AM steering northwest, at noon steering west.

Sunday, January 24th

3:00 PM steering west-southwest—wet hold as usual. At sunset weather looked bad, luffed heading southeast—rainy, squally and miserable during the night. At 5 AM squared the yards steering west by south—this looks suspicious, the Old Man feels decidedly uneasy. Still squally and raining plentifully. Set sea watches (hurrah) now for a passage—catching water and sweeping decks—fine day for a whaleman's Sabbath.

To keep the staves of the oil casks in the hold from shrinking, causing the oil to leak, the casks were carefully watered down periodically. On the *Clara Bell*, this was a job for the boatsteerers. Now that the vessel was no longer cruising for whales, the whaleboat watches were terminated and the vessel went back to the normal sea watches with half the crew on duty alternately. Again they caught rainwater to refill their fresh water casks.

Monday, January 25th

Steering west by south. No improvement in the weather, plenty of wind and rain all night. 12 o'clock midnight took in the main topgallant sail, weather increasing to a gale, heavy seas board us occasionally but still we drive before the wind with full topsails, go it *Clara* I believe you have begun the homeward course. At 4 AM had to double reef the topsails, blowing quite a gale. At 5 AM furled the foresail and fore-topsail close reefed the main top sail and luffed for the purpose of lying to, heading about east by south. Not much rain this morning, wind about the same, heavy seas. At noon set double reefed topsails and foresail.

Tuesday, January 26th
Heading northwest, at 2 PM the wind hauled 'round to the southeast, squared the yards steering northeast, after tea set full topsails. As the weather seems to be improving—still raining some semi-occasionally. 9:30 PM set the main topgallant sail, 6 AM steering north-northwest. At noon steering north rounding Cape St. Mary, bent the main topmast and topgallant staysails.

Wednesday, January 27th
Steering north. Repairing the pumps. Wind south-southeast, set the fore-topgallant sail. Towards evening it was somewhat rainy with a variable wind— squally withal. 7 PM were obliged to double reef the topsails and luffed to the wind heading south. At 12 PM wore ship, heading north-northeast, at 6 AM heading north, furled the fore-topsail. *Lay to* all day, blowing great guns—obliged to pump ship every half hour or so.

In normal service, the pumps might be "tried" once or more a day to remove any water that had accumulated in the bilges in the bottom of the vessel. It is not known if the *Clara Bell* had a pair of simple pumps with plungers to suction water up or more modern chain pumps with small buckets attached to raise water. With the vessel clearly leaking, repairing the pumps was essential. To lay to, the vessel would steer into the wind with just enough forward sails set to counterbalance the rudder so the vessel would keep its bow just off the wind, riding the seas most easily.

Thursday, January 28th
Lay to all night, pumping like good fellows, head northwest, quite rainy with heavy wind. 3 PM wore ship, head east, set full topsails this morning—becketing water casks for going ashore—but where is the shore, 'taint in sight yet.

"Becketing water casks" means making up rope loops to attach the water casks to each other to raft them ashore for refilling.

Friday, January 29th

Heading east by north. Stiff breeze from north-northeast. At sunset it fell almost a dead calm. Main topgallant yard on the cap. 10:30 PM wore ship heading north by west very light winds all night, this morning calm, set the fore-topgallant sail, bent all the staysails. Becketing water casks, heading east for Madagascar, for we have drifted away into the middle of the channel.

Having been blown west into the middle of the Mozambique Channel, the vessel now set all sail to head east to Madagascar. Fore-and-aft staysails could be set from each of the stays between the masts.

Saturday, January 30th

Bent and set the main topmast, main topgallant, and mizzen topmast staysails.

At 10 PM tacked ship heading northeast by north, this morning we have excellent weather with a light breeze, sent up the fore and main royal yards with the sails bent on them already for setting. 8:30 AM wore ship heading east by north a half east, quite warm.

The royal (uppermost) yards were often disconnected and lowered to the deck to simplify the rig while on the whaling grounds or in heavy weather. They could then be hoisted back up and reset when needed.

Sunday, January 31st

Heading east with very light wind. At 4 PM heading northeast, wind hauling to the westward, 5 PM steering north-northeast, wind from south and west light. 7 PM took in the topgallant sails and staysails. At sunset land in sight, or some optical illusion to the eastward of us, which we call land. 11 PM luffed to the wind with the main yard aback, heading southwest, wind quite variable. At 4 AM set every rag of canvas, hardly enough wind to fill the royals—wearing and tacking ship quite often. 9 AM steering north-northeast trying our hardest to beat, tear and wear towards land with little or no wind.

"For Madagascar" depicts the *Clara Bell* under nearly full sail, including gaff topsail, headed for Madagascar to repair the leak. Weir depicts the rig in some detail, with reef points on the courses and topsails, braces from the yardarms, and the various furling lines on the square sails. Under the bowsprit, the martingale runs from the outer end to the foot of the vertical dolphin striker. Aft of that, a martingale guy runs to each side of the bow. Earlicr in the voyage, boatsteerer William Johnson was washed to his death while working on the guys at the foot of the dolphin striker. Weir left off the whaleboats and davits in this view.

Monday, February 1st

Land plain in sight from deck. 2:30 PM stowed away all the iron work, boats craft etc. in order not to tempt the natives. At sunset land between eight and ten miles off—hove to for the night and took in all light sails, heading about south, tacked ship at 12 PM heading east by north, very little breeze all night, calm at daylight—10 AM Begun to breeze up a little, set all sail steering southeast, Sandy Island in sight, washed the starboard side of the ship, ripping up part of the sheathing etc.

"Boats craft" were the harpoons, lances, and cutting spades kept in the whaleboats. Sandy Island, now Nosy Ve, is just south of St. Augustine Bay. The deck planks of whaleships were normally covered with a layer of thin pine planking to prevent wear. Once whaling operations were concluded and the vessel was headed home, this sheathing would be torn up to let the deck planking dry to prevent rot.

Tuesday, February 2nd

We have a fair breeze though light—every rag set that is bent.

At 5 PM dropped our mudhook in Augustine Bay and found ourselves in company with the French brigs, two of which are slavers—had to turn to immediately and break out paint etc. our decks are swarming with natives. This morning set up the bobstays and painted ship outside—rafted 14 casks for water—and now we only await orders from the Prince before we dare to go ashore for anything.

Bobstays, often of chain, ran from the bowsprit down to the bow just above the waterline to counteract the upward pull of the stays on the bowsprit. They could be set up, or tightened, with their sets of deadeyes.

Wednesday, February 3rd

Painted ship, scrubbed decks. Raft banging away astern, Prince Millie not having given permission to get water yet. Took the casks on deck again.

As Captain Robbins explained in *The Gam* (see note for February 28, 1857), the head men representing the bay and the upriver populations had to agree and be compensated before a crew could land and refill the water casks. The casks were roped together into a raft that would be towed upriver by the whaleboats to be filled from a clear-running stream. As fresh water is lighter than salt water, the full casks remained afloat as they were towed back to the vessel.

Thursday, February 4th

Saw His Highness the Prince, took the raft up the river, filled the casks. At 4 o'clock AM started for the raft, had it alongside by 7:30 AM and breakfasted, hoisted the water on deck and rolled it forward, in order to bury the ships head and lift the stern to get at the leak. Chips has been up to his armpits in the water working at the leak all the morning. A sail in sight coming in. Went ashore with the captain to get a bullock and a time we had too—two tribes of the natives were fighting, popping away at one another with their shooting things. When they had popped off about a keg of powder one party retreated to the hills, leaving one dead man upon the field.

By the 1800s, large ocean-going vessels commonly had a layer of thin copper plates nailed to their bottoms to prevent boring "shipworms" or teredos—actually a form of clam—from riddling their planks and keel and to limit the growth of marine plants on the bottom. To find the leak, the crew had to put the filled water casks forward to raise the stern slightly. Then Robbins and "Chips"—Carpenter William Campbell—investigated the stern closely and discovered that the copper sheathing was not properly soldered along the sternpost. Even that small seam exposed the planking to teredos. "Through the tiny water-course thus left exposed, the worms had got in—puny creatures, so small you could barely see them with the naked eye, yet they bored those solid white-oak planks into a sieve like a honey-comb," wrote Robbins. "A blow from a hammer on the worm-eaten parts would crush the wood in like an egg-shell. There was a weak spot, easily repaired that might have sent us to Davy Jones" (Robbins, *The Gam*, 202–3).

Friday, February 5th

Our old friend the *John Dawson* hove in sight, came in and anchored near us. 5 PM the carpenter had finished repairing the lead, stowed most of the water in the blubber room, leaving four or five casks to be stowed forward or left on deck.

Saturday, February 6th

This morning the starboard watch went ashore for a run, and whatever else they could get—could not enjoy ourselves much, took pencil and paper ashore but could do but little. Larboard watch stowed the water and cleared up decks.

Sunday, February 7th

Had quite a severe blow off shore this evening, had to stand in readiness to drop the starboard anchor, this morning the larboard watch went ashore for Liberty—something or another to be done all day long—the Commodore being aboard.

The *Clara Bell* had anchored in St. Augustine Bay with the larboard "small bower" anchor. With the wind rising, they prepared to drop the starboard "best bower" as well to prevent dragging anchor and drifting ashore.

Monday, February 8th

Received half a bullock from the *John Dawson*, all ready for sea and to start direct for home. At 8:30 AM weighed anchor from this delectable spot, and now we are homeward bound. We have a moderate breeze, with which we can knock out six knots with staysails and royals; by noon it was quite calm. Put the boats in whaling order but nobody wants to see a single spout from here, home.

Tuesday, February 9th

Bent the studding sails. Steering west-southwest. 5 PM splendid breeze from the south, 3 o'clock morning watch set the larboard fore topmast studding sail, this forenoon set the larboard and starboard main-top-gallant studding sails—this looks something like, what a mass of canvas, but we can't go fast enough towards home.

Wednesday, February 10th

Set the lower studding sails, hurrah old man crack on the sail as you have always said you would do so soon as our good ships head was pointed for home. We now steer west-southwest with a pretty fair breeze from the eastward—it looked squally after sunset and we took in the royals, but they were set during the midnight watch.

The more sail we keep set, so much the more lively do the men become—the general topic is, home.

The *Clara Bell* set studdingsails whenever the wind allowed. Here, Weir shows the vessel sailing downwind with fore topsail and fore course studdingsails set to starboard and main topgallant studdingsail set to larboard. Studdingsails had a loose boom at the head that was suspended from a block on a yardarm, and studdingsail booms were extended out from the yardarms below to secure the clews of the studdingsails.

Thursday, February 11th
Still bowing along with a seven knot breeze—studding sails and royals free, course west-southwest. Commenced ripping the sheathing from the deck, and most of the watch are employed in scraping up the dirt and tar which has collected between the sheathing and deck.

Friday, February 13th
Ripping up the sheathing and with the best boards Chips has boxed in the tryworks. Steering southwest by west a half west with wind from the east and north, bringing fine weather.

In addition to removing the deck sheathing, the tryworks was also decommissioned. Usually it was dismantled and the bricks were thrown overboard, with the trypots saved for the next voyage. In this case, Weir implies that they built a box around the tryworks.

Saturday, February 13th
Steering southwest by west a half west, scraping the decks and plugging up nail holes.

At sunset the weather looked very threatening. At 10 PM we're obliged to take in studding sails alow and aloft royals and topgallant sails, wind quite variable with little rain, just enough to make it villainously unpleasant. 11:30 PM midnight watch braced sharp on the larboard tack, heading west by north, the wind comes directly from the point which we wish to steer to.

Shipped in the studding booms and unrove the rigging—all hands in a rather growly fit. 5 AM double reefed the fore topsail, at 6:30 AM double reefed the main top sail, heading west by north. At noon prospects begin to brighten—shook out the reefs set the main topgallant sail with all the fore and aft sails bent—head west.

Sunday, February 14th
Weather looks decidedly bad, we head anywhere between west and west-southwest, at day light set the royals, heading west by south, 9 AM tacked ship heading southeast by south—raining some.

Monday, February 15th

Awfully dull weather, heading southeast a half south, nearly a dozen points from where we want to steer to. Enough to make a deacon swear. 4:30 PM took in the topgallant sails; 9:30 PM double reefed the topsails—oh my! shall we ever reach home? Shall we ever round the Cape and be able to say that we are in the Broad Atlantic—for that is next door to home—say shall we ever do this, at the abominable rate we are now running. 10 PM wore ship heading west by north—a heavy sea on, but our bark rides it splendidly. 4 AM wore ship heading south-southeast, wind quite variable, 6 AM wore ship, heading west, strong winds from the southward to westward. Sore heads plentiful else what do we all growl so for, cause we can't go straight home.

Tuesday, February 16th

Heading west by south—weather improving, 6 PM set main topgallant sail, 10 PM royals and staysails all set, with a smacking breeze, steering southwest—hurrah—this does look like going to swate Ameriky, wind southeast. By daybreak wind from northeast, moderate, squared the yards and set the studding sails alow and aloft, overhauled the main spencer rail blocks. By noon the wind was quite fresh, and our drooping spirits have revived, hands employed in making seizing stuff.

Weir's "swate Ameriky" parodies the Irish immigrant's pronunciation of sweet America. The main spencer rail blocks were the blocks on the rails through which ran the sheets of the fore-and-aft spencer. Seizing stuff is small line used for wrapping or securing.

Wednesday, February 17th

Steering southwest. Studding sails and royals a clever full, 2 PM steering west-southwest, from which time we had a stiff breeze till after sunset, wind seems to be hauling to the northward. 10 PM blowing quite fresh, took in studding sails and royals. 10 PM wind from northwest, heading west-southwest by noon stiff breeze, heading west by south.

Tuesday, February 18th

Stiff breeze from the north and west, heading west, sailors making spun yarn and they do rattle that organ most unmercifully. 4 PM set the top-gallant sails, 9 PM we were heading the course west-northwest, wind somewhat variable. A great deal like that last variation nothing definite about it.

This morning all light sails set close hauled to the wind, heading west, sailors splicing yarns etc. 11 AM wind shifted to the southward, light, braced round the yards heading west. At noon we have what is called a jolly Scotch mist—wet a man through in ten minutes—aye—in four minutes.

Weir had not visited Scotland, but he knew the common term "Scotch mist" for a mixture of fog and drizzle or light rain.

Friday, February 19th

Breeze freshening. 2 PM heading northwest, 4 PM took in the light sails, heading northwest by west a half west wind a little variable and squally during the night. 6 AM double reefed the fore top-sail, 8 AM shook out the reefs—rainy all the forenoon, a doubly double Scotch mist.

Saturday, February 20th

Very squally. At 2 PM double reefed the topsails, heading southwest.

3 PM, a heavy squall struck us, hauled up the courses, dropped the topsails took in jib and spanker and double reefed in short order. Wind knocked us off northwest—more or less, 4:30 PM furled the fore top-sail, 6 PM wore ship heading south-southwest, quite squally and rough. 5 AM wore ship, heading north-northwest, set full topsails. 10 AM wore ship heading west. Noon heading southwest, broke out water and flour, we should have had bean soup today for it is quite rough enough.

Sunday, February 21st

Under double-reefed topsails, heading southwest. 2 PM tacked ship, 3 PM had the fore-top-gallant-sail set, heading northwest and north

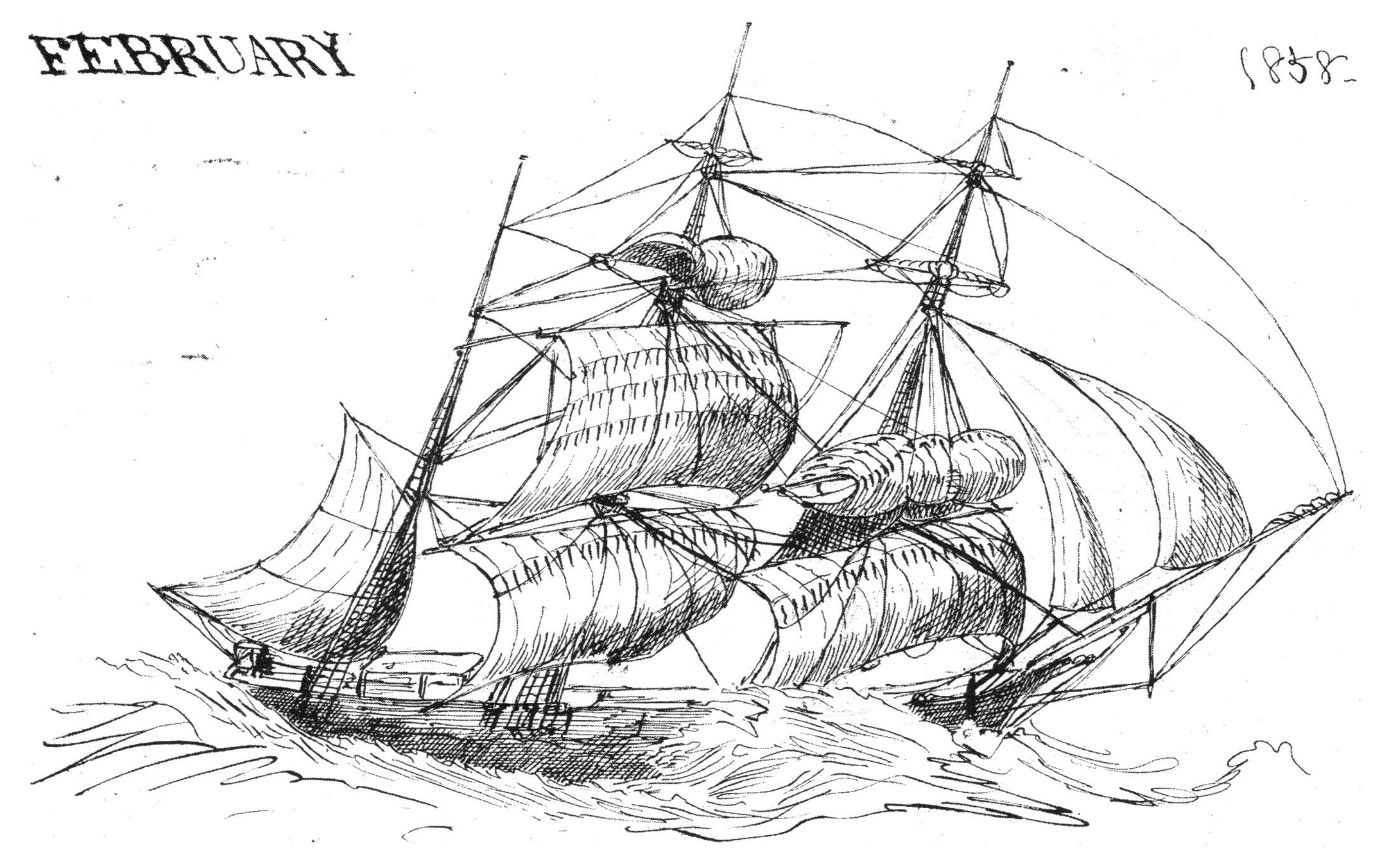

In heavy seas, the *Clara Bell* runs under spanker, main course and topsail with topgallant clewed up, fore course and clewed-up topsail, and inner and outer jibs.

northwest, 4 o'clock set the royals, 6 PM steering west-northwest right straight for home—but how long will the wind allow us to continue so? At 1 AM quite calm, towards daylight had variable winds, at 6 AM wind rather light but fair from the east, set studding sails alow and aloft—but we can't make over four knots with this wind.

Monday, February 22nd

Steering west-northwest, by sunset quite calm, with a heavy swell on, the sails make a horrible noise flapping against the masts—blow! Good breezes blow! 3 AM wind shifted to the westward almost contrary to what we would wish. At 6 AM had a strong breeze; tacked ship, head north by west, 9 AM wind increased to a gale, had the topsails double reefed, heading north, at noon blowing great guns, furled the fore-topsail and close-reefed the main—sail in sight.

Tuesday, February 23rd

Heading north with a stiff breeze and rugged sea. 5 PM heading north-northwest. 9 PM set the topgallant sails, wind light but fair, steering west by north, at 10 o'clock dead calm, dropped the yards on the caps, for of what use is duck now, with the wind right up and down—surely we won't approach America this night, but have patience my boy.

At daylight we have a light breeze from southwest, and now we have *crowded* on every rag of duck that there is room for on our spars—

Two sails in sight supposed to be merchantmen. 7 AM steering west-northwest, with a fresh royal breeze.

Caps are the short fore-and-aft timbers at the top of the overlaps between sections of mast. When lowered, the yards hung from the lifts that connected their yardarms to the mast.

Wednesday, February 24th

Hurrah! the wind is increasing, and we are thankful—though it is not quite fair for our course, west-northwest—three sails in sight.

At 2 PM a heavy squall struck us, everything was *let go* by the run—topsails were clewed up and shivering in a trice, the spilling lines, clewlines, slab lines and buntlines of the mainsail were manned double quick—such is the moment when a sailor is all life and animation. The ship with her lee rail even with the water, wind howling through the rigging and spars, ropes ratting, and the sea roaring—it is glorious. After the squall was over, set all sail and broke out a new fore topsail, for the squall blew a hole in the old one. 4 PM sent down the old fore top sail and bent the new one. At sunset the wind began to die away, at 10 PM heading north by west, 11 PM tacked heading southwest, 2 AM breeze freshening, 6 AM tacked ship heading southwest by west, four sails seen. At noon, it blew quite fresh—took in all light sails, heading west-southwest.

Weir mentions the various lines used in furling the courses and topsails. Back on December 16, he mentioned replacing most of these lines.

Homeward Bound—February 1858

Thursday February 25th
Stiff breeze from the northwest. Three sails in sight, all bound for the Atlantic. 3 PM getting squally—double reefed the top sails and furled the main sail, jib and spanker. Heading west-southwest, with heavy sea. 4:30 PM gale freshening, furled the fore topsail heading west a half north. 5:30 PM blowing a perfect gale laying to, under close-reefed main topsail and fore top mast staysail, the hatches are barred down, and all preparations made for a bad time, our noble bark rides well, much to the mortification (no doubt) of an English merchantman who is laboring heavily on our lee. 11 PM furled the mainsail and set the fore and main spencers, blowing great guns. At 7 o'clock this morning it cleared off finely, allowing us to set topgallant sails, though the sea is still quite rough.

Friday, February 26th
Weather looks doubtful, we are now under double reefed topsails and courses. 4 PM set topgallant sails, heading northwest, one sail in sight 11 PM tacked ship heading west, at 12 PM steering northwest with good weather—set studding sails alow and aloft, with royals. This morning two sails in sight, by noon stiff breeze from the north and east which will just allow us to head a course. I wonder if those at home can imagine where I am now, or have they ceased to think of me. If I could get a single line from my father I would feel content—let it censure me ever so highly, it would give relief.

Saturday, February 27th
Heading northwest. At 1 PM took in the light sails—three sails in sight. At 3:30 PM the jib sheet parted, and every flap of the heavy duck sounded like the report of a cannon, we hauled down the sail and furled it, and took in the spanker and mainsail. 4 PM furled the fore-topsail it bids fair for another gale, head south-southwest 2:30 AM wore ship heading north-northeast parted the storm staysail pennants—temporary ones were rigged with preventer sheets.

This morning at daylight we made new staysail pennants and bent them, the foresail is furled and main topsail close reefed—gale increases fast.

A sheet is the line attached to the bottom corner or corners of a sail to pull it taut while setting. Jib sheets might be eased or taken in, depending on wind direction, and suffered a good deal of strain. A storm staysail was a small triangular sail set on the forestay to maintain steerageway in very heavy weather. The storm staysail pennant was a line to secure the tack, or lower forward corner, of the sail to the bow. Preventer means backup, or additional.

Sunday, February 28th

At 4:30 PM the gale had abated somewhat, so that we set a close-reefed fore-topsail, heading northwest, one sail in sight. At 6 PM set the foresail, 2 AM wore ship heading southwest. The weather is growing better slowly, and we make sail accordingly, by daylight had topgallant sails and staysails set—steering north-northwest, three sails in sight. It is quite cold, and the night watches are anything but pleasant here.

Monday, March 1st

Time wears apace.

Set the foretopmast studding sail, steering north-northwest. Wind from west-southwest, before 3 PM set the lower larboard studding sails and the main topgallant studding sails. One sail in sight.

6 PM set the royals, moderate breeze, steering north by west, wind from southwest. This morning we have good cool weather and on the strength of it we broke out flour and water, the sailors are at work making rope yarn etc.

Tuesday, March 2nd

Working miscellaneously, here there and everywhere, and the boys growl though the wind is fair. Packed away the old topsail and jib and the mate says the Old Man is going to speculate with them when he gets home,

"Small Fry" shows two boats amid a school of blackfish. The boats took five blackfish, though one of them poked its melon (forehead) through Weir's boat.

sell them at 4 cents per pound to the paper mills. Fine weather but little breeze. 10 AM raised a large school of blackfish, lowered three boats and got in the very midst of them. I threw the iron into a big fellow when he turned about and poked his nose through our bottom, the men put two or three shirts in the hole and we fought and killed Mr. blackfish, the waist boat got three, and the starboard boat with the Captain and Mr. Perry got two.

Weir alludes to the fact that, at that time, most paper was made from rags reduced to a fibrous slurry, then screened and dried at a paper mill.

Wednesday, March 3rd

By 1 PM all boats crews were aboard, and the five fish hoisted on deck. I wonder what a backwoodsman would say at such a sight—five great blackfish, not one of which would weigh less than two tons. But now comes the disagreeable operation of flaying them, a hard looking mess; all hands employed about the fish are soon thickly besmeared with blood; and it is with difficulty one can stand up in the clotted gore—talk about the horrible sights of a slaughter house, and then look here, a single glance at our decks will do.

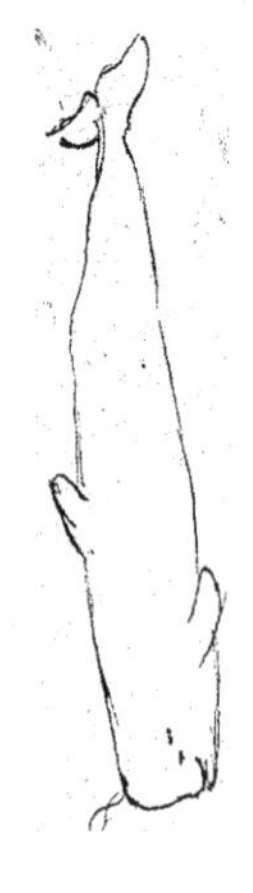

"Stripping Blubber Off Blackfish Whales" shows the processing of these small whales on deck. "Talk about the horrible sights of a slaughter house, and then look here," remarked Weir with revulsion about the process. Although it was far bloodier to cut in a large whale, it was done in the water and the men were spared much of the gore.

We have a light southwestly breeze with every rag set. Steering north but we cannot make more than one knot an hour—blow good breeze blow! Hacking away at these mammals till 4 PM. The weather is beautiful and the ocean as smooth as a river, but this won't take us home. The sailors growled some because we stopped to take these fish, and make such a mess, for which, ten to one none of the crew will be paid.

6 PM hauled aback the main yard and passed the larboard boat on the starboard side for repairs, breeze freshening, so that we had to take in studding sails though not without growling on the part of the boys.

Hang the Blackfish the *Old Man* makes more bother about them than he would over an 100 bbl whale, and they will not make us over six bbls at the most. This morning the weather is dull and we all feel the influence of it, the blubber is minced and ready to try out.

Thursday, March 4th

Started the fires, and by 4:30 PM the blubber was tried out and yielded us six and a half bbls, who'll grow rich on that speculation, not I, I'll wager my old hat.

6 PM we had finished repairing the damage done to our boat, and set her on the starboard boat's cranes and the starboard boat on our cranes—so now all three boats have their patched sides towards the ship.

At 7 PM steering north by west, with a fine breeze, studding sails and royals set, and we move merrily towards home, home home home. By 8 AM steering north-northwest, breeze freshening.

Friday, March 5th

Steering north-northwest with fine breeze. Sailors hand organ alias the spun yarn machine in operation—cheer up boys, your bondage will soon be ended. 6 PM, blowing a smacking breeze, parted the seizing of the larboard fore topsail, brace block; rigged up a temporary one. 9 PM wind still on the increase, had to take in studding sails. 11:30 PM it again moderate so we piled on the rags. This morning the weather is doubtful repaired the brace blocks and if they do not last from here to home in any kind of breeze I'll heave my old hat over.

Lat 25° S, Long 2°7'

Saturday, March 6th

Steering northwest by north. With a stiff breeze and all sail set, and above all bound for home. Bent new spanker gear this evening; we are priming up everything about this time for the prospect is we shall soon sight St. Helena; where we shall in all probability paint scrape and wash the old ship from truck to keelson. At sunset the weather looked bad, but so long as the wind is fair, I say blow San Antonio blow and waft us to the other edge of the Atlantic.

This morning the sailors are at work upon the lower fore rigging preparatory to turning it in and rattling down. We have had a fair breeze all this week, and no one remembers our fancy bark to have been driven so long a time before with favorable winds. We are thankful. Lat 23°58' S. Slow traveling but we know not how sure.

Sunday, March 7th

Steering northwest by north, breeze fresh and fair. Hands employed scraping the pitch from the decks—wet hold as usual. Had a splendid breeze all night and this morning the weather is beautiful for the day, as it should be. It is hardly 8 o'clock with us, but with the good people at home it cannot be more than 3 in the morning, surely there is no one thinking of me now—but they may be dreaming. Oh! what a splendid day it is, if I could only walk down to the little church with Em this morning I would be happy—there I am growling discontent, I'm fast getting to be an old salt, or getting salted.

Monday, March 8th

Steering northwest by north. Fresh breeze continues, and we are more grateful than ever for it. 3:30 PM. They are in church now at home and I wish I could be with them. Can I ever go to that dear little church again, can I ever see my dear father, my brothers and sisters but where shall I look for an answer, still I feel that I shall be numbered among all the dear ones again, but it is wrong for me to think anything about it—God's will be done, as always.

Lat 21°19' S

Tuesday, March 9th

Steering the same course northwest by west. Moderate breeze from the south, the weather is somewhat cloudy and threatening. Rattling down and serving the fore rigging, scraping decks, planed the spare spar, and cleaning up things for port. Long 1°23' W, Lat 19°50' S

Wednesday, March 10th

Steering north-northwest. Excellent breeze continues: finished rattling down the fore rigging, the boys scraped all the masts. Prepared soap and water for washing the paint work, oh! My such a house-cleaning as we'll have shortly.

A fine smacking royal breeze from southeast blowing all night—so far good. This morning commenced the operation of washing the paint work inside the ship—no house wife ever saw such scouring, and what is more I trust they never will. Hurrah fair St. Helena—Lat 18°31' Drive on *Clara*, my pretty bark; you seem to be inspired at the idea of going home too.

Thursday, March 11th

Three sails in sight at sunset, one of which the Skipper thinks is a whaler. Steering north by west, kept off west for one of the sails, and at 8:30 PM though it was quite dark, we spoke and gammed with the bark *R. L. Barstow* 700 bbls sperm. No news of import. 9:30 PM took in topgallant sails and studding sails and double reefed the fore topsail. 1 AM hauled aback the main yards and waited for daylight for the old man says land is not far off. By daylight we were startled by the joyful cry of "Land ho!" Made sail for it with a stiff breeze. By noon the mudhook will be dropped off Jamestown, St. Helena—then hurrah for Liberty. At 6 AM all hands on deck cables hauled up, anchors loosed and ready for dropping. 7 AM we are bearing towards land bravely, and all hearts are gay at the prospect, for here we shall meet friends in a civilized port. 10 AM let go the anchor, but the cable was not long enough to drag and catch and we drifted away to leeward of the shipping feeling not a little mortified at such a proceeding. It was blowing quite fresh and we beat up to the anchorage under double reefs.

Named for the Mattapoisett shipbuilder and whaling agent Robert L. Barstow (owner of the *Clara Bell*) and launched in 1851, the 203-ton bark *R. L. Barstow*, commanded by Captain Edward Davoll, departed for the South Atlantic whaling grounds on August 19, 1856, and returned on August 30, 1858, with 704 barrels of sperm whale oil and 115 barrels of right whale oil. Three years later, Weir would serve under Captain Davoll on the whaling schooner *Palmyra*.

Friday, March 12th

A few minutes before 1 PM we dropped both anchors inside of the larger shipping and this time the mud hooks stuck.

There are 28 vessels at anchor here now—English and French muster the strongest. The US Sloop of War *J. Q. Adams*, is anchored under the lower fort, and already there has been a number her officers on board of us inquiring for shells, curiosities etc. etc.

Here, Weir misnames the US Navy frigate named after President John Adams for his son President John Quincy Adams. Launched at Charleston, South Carolina, in 1799, the *John Adams* served against the Barbary corsairs in the Mediterranean but was laid up through the War of 1812. After diplomatic duties and expeditions against piracy, the *John Adams* was rebuilt in 1829–1830. It then carried out further diplomatic and convoy duty and a round-the-world cruise before being laid up, except for brief duty during the Mexican–American War. Assigned to patrol for slavers on the Africa Station from 1849 to 1853, the *John Adams* was probably returning to its subsequent Pacific Ocean duties when encountered at St. Helena.

Friday, March 12th—St. Helena

The captain has gone ashore and the mate is gamming, but we poor sailor men and slewers are left to clean ship, send down the sails and get everything in readiness for painting. No one will be likely to overexert himself this day, for the first day in port is generally considered a holiday, therefore though we've got work on hand, it is done but slowly, and I care not a fig if it is not done at all. I shall go and stow myself somewhere in the rigging and have a jolly old smoke, feast my eyes on the terra firma sights, and my lungs upon this delicious breeze directly off shore.

Saturday, March 13th

What a beautiful day, but I cannot enjoy it, here must I remain while the other watch goes ashore for liberty. Three more sails have arrived today, and some five or six weighed anchor.

It is right pleasant to hear the chimes which the different bells on board the ships make all the hours of the night are struck, from sunset to sunrise, and such a variety of vessels produce an equal variety of tones, and hear bells striking one after another, make pleasant music to the watchers' ear.

Sunday, March 14th

Two of our sailors created quite an excitement during the night. Some of the watch below were aroused by a violent scuffling on deck, and when on coming from the forecastle they saw two of the boys fighting, they immediately attempted to part them, but it was too late, for our yankee lad had received a severe stab, which prostrated him.

The mate was on board fortunately, and was on hand to attend to the matter. I had to turn out and take a boats crew to the *John Adams*, hail the sentinel, and get a surgeon, which was done with little difficulty. The surgeon came, and when he had probed the wound and bound it, the sailors picked the boy up to take him to his berth, when our man o'war's surgeon says handle him gingerly boys, in such forced nautical style as to cause a broad grin on our older sailors countenances. It told immediately he had not lived long upon the sea. The other boy a Gee is now handcuffed and placed in the run, enjoying his hard bread and water.

This morning the English frigate *Sybel* arrived from the coast. And the noise of their salutes, combined with the guns of the fort and of the *J. Adams*, sounded like music to my ears, for it savored of the old Point more than anything yet heard or seen.

A short time before we came in here, that famous slaver *Windward* was captured by H.B.M.S.S. *Alecto*. Nine and a half hours after sighting the *Windward* the *Alecto* overhauled her—the sea was rather calm or it could not have been done so readily. The hull of the *Windward* is now plainly seen from our decks. She is anchored close by the shore, there to be left to rot. And oh! What a beautiful model it is, the most perfect clipper I feel almost like crying, while looking at her and thinking that it is a loss to all.

This stabbing was more evidence of the friction between "Yankee" and "Portuguese" crew members. As before, the guilty man was placed among the casks in the run under the officers' quarters. Both the British Royal Navy and the US Navy patrolled off the African coast to suppress the slave trade that was still carried out, especially to Cuba. United States participation had been outlawed since

1800, but some American vessels, including a few whaling vessels, attempted to profit in the illicit trade. St. Helena was the closest British territory to the African coast and, beginning in 1840, served as a Royal Navy station where enslavers faced an admiralty court and the enslaved were held for return to Africa. HMS *Sybille* was a 160-foot, 36-gun sloop of war launched in 1847. The 164-foot, steam-powered sidewheel sloop of war HMS *Alecto* was launched at Portsmouth, England, in 1839. In 1845, the *Alecto* engaged in three tugs-of-war with the screw steamer HMS *Rattler* that demonstrated that the screw propeller was more powerful than paddle wheels. Beginning in 1852, the *Alecto* was assigned to anti-slavery duty off the west coast of Africa. During its fourth deployment, 1855–1859, the *Alecto* seized five slaving vessels, including the swift brigantine *Windward* of New Orleans. The *Windward* had gone to the Congo River in October 1857 to load enslaved Africans, probably for delivery to Cuba. Evading US Navy surveillance, the vessel took on 603 Africans and headed to sea. After a 12-hour chase, the *Alecto* seized the *Windward* on November 4, 1857. By the time they arrived at St. Helena on November 16, 149 of the Africans had died. On May 1, 1858, the *Illustrated London News* would publish a wood engraving titled "Night Chase of the Brigantine Slaver 'Windward' by HM Steam-Sloop 'Alecto.'"

Monday, March 15th

Went ashore this morning and the first place I steered for was the market, for I longed to get a bite from a peach again and here there are peaches in abundance. I purchased what grapes, peaches, pears and figs I could well carry in a handkerchief, and Holt, one of our sailors who agreed to accompany me, filled his handkerchief with apples etc. And off we started for Napoleon's tomb and Longwood. The scenery is glorious it reminds me of what I have heard about the Alps. The hills are rocky and steep, and the paths are cut on the very edge of deep chasms and high precipices, and occasionally we would meet a long string of donkeys laden with produce for the market, wood etc.

By noon we were at the Tomb, and a most beautiful spot it is, at the bottom of pleasant vale and nearby is a spring of the most delicious cool water—here we sat down to rest after our weary climb, and eat the fruit which we had brought for our dinner, after which we smoked and chatted till we thought it time to leave for the ship and by sundown we hailed the boat and they took us on board with what fruit we could carry.

C. E. Holt had joined the *Clara Bell* as a greenhand, like Weir. Longwood was about seven miles from, and 1,600 feet above, Jamestown. A 40-room, one-story house, Longwood had been converted from a farmhouse to serve as Napoleon's home in exile. He asked to be buried in the Valley of the Geraniums, where he had walked, and near the spring from which his water had been brought. His body lay in a simple tomb there from his death in May 1821 until October 1840, when his remains were returned to Paris. The French government purchased Longwood and the site of the tomb in 1858.

Tuesday, March 16th

Today the larboard watch are off on liberty, and we are engaged in painting the yards and masts. All the sails were unbent yesterday, and left hanging by the earrings, and all running rigging has been cleared for the painters.

The superintendence of this work is left to me as usual, and I make a great mystery of it—to make certain colors I have used as many as eight kinds of paint when the requisite tint might have been obtained by mixing only two colors. However it is all in the three years. The color we are now to paint the interior of the bulwarks with care can be made from burnt sienna and white, with a very little chrome yellow; but I had to mystify the old skipper and mate by turning 50 pounds of white paint into a tub, then four pounds of raw umber, one pound of yellow ochre, one pound of red lead, a sprinkle of vermillion, a touch of kings yellow and burnt Sienna. So here we have about 60 pounds of splendid coloring for our noble bark's bulwarks, but a little exposure will change this magnificent tint materially—it will last till we get home, I am confident. If it don't, we already know the paint is very bad, a spurious article.

To paint the yards, the sailors cut the lashings securing the heads of the sails to the yards, leaving the sails hanging from only the lines called earrings that secured the upper corners to the yardarms.

Wednesday, March 17th

6 AM waiting for the other watch to come on board, heard they had a row ashore last night.

Thursday, March 18th

Went ashore yesterday and crossed to the other side of the island, coming back we tried to make a bee line for St. James town, but missed our path, got benighted, and had to encamp out till morning. Fortunately, the weather was very favorable for the situation we were placed in. We built a hut of palm leaves, kindled a fire and roasted a few bananas for supper, after which we smoked and talked about the good people at home till we fell asleep. We were up bright and early in the morning, went to the marketplace, and made a breakfast off of fresh figs and grapes topped with a roll and some splendid coffee, the like of which It seemed we had never tasted before, and by sunrise we were on board the ship, ready for labor. Painted the outside of the ship and forward the foremast to the eyes. I varnished the cabin, a good day's work, as my poor wrist can well testify, for it feels somewhat achy with the exertion.

Friday, March 19th

An unpleasant day it has been, but I have been ashore this morning and brought back about a bushel of peaches and pears for we sail today.

The sailors have crowded so much fruit into the forecastle that it looks more like a fruit venders stall, than any part of a ship. I went down into the steerage for pipe and tobacco and found that place crowded with bags and baskets of all the kinds of fruit which the island afforded. Some of the sailors finding my chest filled have thrown over a bushel of bananas and plantains in my berth a fruit of which we have seen so much that we care little about, though any fruit is prized at sea for one soon tires of salt junk.

Saturday, March 20th

Weighed anchor from St. Helena between 3 and 4 o'clock PM. The whale ship *Almira* started about an hour before us and lay to in the offing waiting. It is blowing a smacking topgallant breeze. Steering north-northwest in company with the *Almira*, as it was cloudy during the night, we ran under short sail, and kept a light hung in the mizzen rigging all night.

This morning the *Almira* is off our larboard beam within half a mile.

The *Clara Bell* had encountered the *Almira* near the beginning of the voyage. Now, on its return from the Pacific, the *Almira* appeared again and attempted to sail home in company with the *Clara Bell*.

Sunday, March 21st

Pleasant breeze from the south and east, beautiful weather. Ten sails in sight, set the royals and topmast and lower studding sails. Our next port is home—the haven where we would be, and yet how long will it take us to reach that happy land, we have got the equator to cross and the doldrums, and the chances are ten to one that we suffer from calms, but I suppose it will all be right.

The "doldrums," or Inter-Tropical Convergence Zone, extends about five degrees north and south of the equator and has little wind because the northwest trade winds of the southern hemisphere meet the southwest trade winds of the northern hemisphere there and the heated air rises rather than blowing across the surface as steady wind.

Monday, March 22nd

Steering north-northwest with fair breeze from the southeast. At about 1 PM gammed with our consort, at 2 PM spoke the *H. B. Milam*, merchantman of New York, bound directly home from China—asked them to report. 6 PM sunset finished gamming, set the royals and topgallant studding sails. By midnight the *Al* was far out of sight astern. Eight sails in sight this morning. One large Frenchman set his challenge and beat us all, but he carried many more sails than we. Repairing the rigging—got the boats ready in case we should see whales. Hang it!!

The *H. B. Milam* is probably the 865-ton, 166-foot ship *H. B. Mildmay*, launched in 1856. The "*Al*" was the *Almira*.

Tuesday, March 23rd

Sailors kept as busy as ever. Put new seizings on the lower fore rigging, beginning to overhaul all the rigging and fit ourselves for home.

At sunset there were ten sails in sight—all bound the same course as we, there are not many of them that can keep up with us. One Yankee and a Frenchman have passed us, but they carry much more duck than we. Steering north-northwest with moderate breeze, during the night we gained considerable on the Frenchman as he is now abreast of us, not over two miles off. Served the fore-topmast backstay—seven sails in sight and five of these will be left out of sight astern of us by night—Frenchy has gained a little upon us since day light.

Lat 11°15'

Wednesday, March 24th

Sailorizing extensively—served the fore-topgallant backstays, etc. etc. etc., steering north-northwest. The Frenchman gained upon us till sunset when we commenced overhauling him again, had a light and variable breeze during the night, by daylight Frenchy is not more than half a mile ahead of us. We can see but four sails now—the remaining six we have left far astern and Johnnie Crapeau would fare the same if we could have a right stiff breeze. Set up the jib and flying jib stays and tarred them, served the main topgallant back stays, and did it after a true sailor fashion—the Old Man will have our bark in excellent order for home. Lat 9°40' S

"Johnnie Crapeau" is a disparaging reference to the French, including the French word for toad as an allusion to the French preference for eating frog legs.

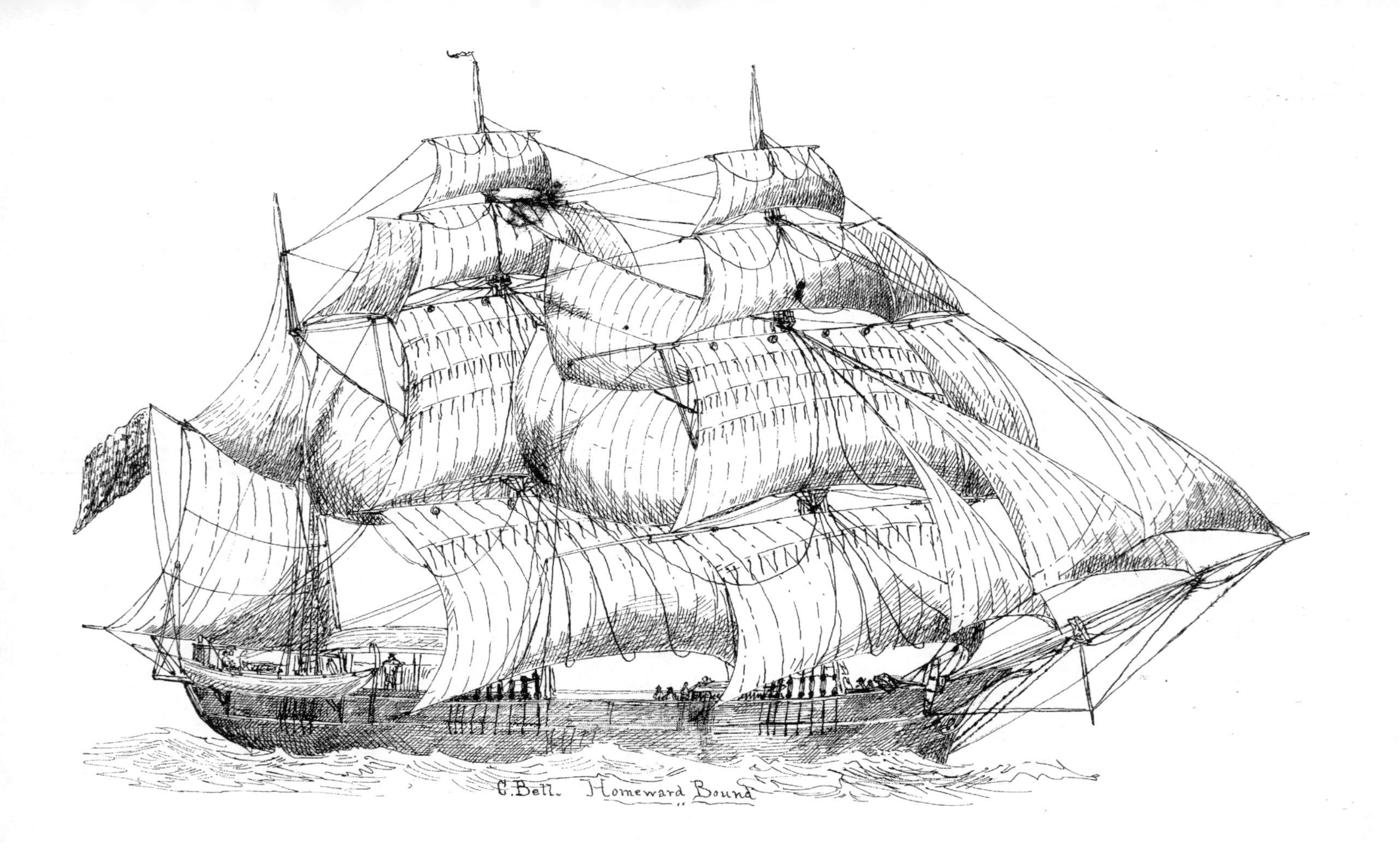
C. Bell. Homeward Bound

Facing page: "*C. Bell* Homeward Bound" depicts the vessel racing home with a favorable breeze. All square sails are set, plus main topgallant studdingsails and fore topsail and topgallant studdingsails. Studdingsail booms have been run out on the fore course and fore topsail yards, but the sails are not set. Fore-and-aft sails include three jibs, main topgallant and topmast staysails, mizzen staysail, and spanker and gaff topsail. Weir includes many details: the bowsprit footropes looped below it for the crew to stand on when handling the jibs; the chain bob-stays securing the bowsprit to the hull, running down and aft from the dolphin striker to the bow, between the martingale guys; a suggestion of the "winged dragon" figurehead; the best bower anchor secured on the rail with its wooden stock standing vertical; the iron chainplates that anchor the deadeyes for the shrouds to the hull below the channels at each mast; the coiled gaskets along the head of the topsails, which would be uncoiled and then wrapped around the rolled topsails to furl them; the captain's starboard boat on its cranes at the davits, with other boats on the overhead skids between main and mizzen masts; and the helmsman at the wheel, just forward of the small house at the stern.

Thursday, March 25th

Set up the fore-topmast and fore stays. Little or no wind, sailors growl accordingly. The Frenchman is right abeam of us this morning—this is pretty even sailing. The copper on our bottom is rough and in bad condition, otherwise, we might leave Frenchy alone and far astern. At 10 AM land in sight "Isle of Ascension" very light winds from southward, can't make 3 knots an hour, by noon our rival is off the larboard quarter—a decided beat.

Lat 8°22' S

Ascension Island lies about 800 miles northwest of St. Helena, 1,000 miles west of Africa, and 1,400 miles east of South America in the South Atlantic. Named by Portuguese mariners on Ascension Day in May 1503, it was visited by mariners for water and green turtles, but it was not settled until the British Royal Navy garrisoned it in 1815 after Napoleon was exiled on St. Helena. An arid volcanic cone, Ascension had little vegetation until the British began to plant introduced species there in the 1850s.

Friday, March 26th
Served the fore-topmast backstays, set up the bobstays and set taught the bowsprit gear. Ascension in sight till dark. This forenoon steering northwest by west. Lat 7°23' S

Saturday, March 27th
Steering northwest by north, abominable rainy weather, a Scotch mist prevailing with very little wind—still we are kept busy upon the rigging.

4 PM four sails in sight, one of which is our French rival, overhauling us. During the night the rain fell in torrents—wind variable, baffling and light. At noon it cleared off beautifully, one sail can be seen from deck, and it is that villainous old Frenchman. Lat 6°21' S, Long 17° W, slow traveling.

Sunday, March 28th
Quite squally around—five sails in sight—and our French friend is not. Broke out water for the scuttle butt and wet hold, what a delectable employment. I can never wet hold without getting half drowned, while the other harpooners keep quite dry apparently, however it is not so very uncomfortable in these latitudes after all, and any way we will not have much more of it to do. Had a very pleasant watch on deck last night—it was moon-light, and a fine royal breeze blowing right straight home. Though it is hardly so strong as we would wish, we are well contented with it. This morning four sails in sight, one Englishman, six miles astern of us by day-light, was abreast of us by half past 11. She seems to go two miles to our one, though we have every stitch of canvas set, she carries a dozen more sails than we and does not appear to be very heavily laden. When we can just stagger under topsails, bring on any Englishman, and we'll beat him, or if we don't they'll have something well worth boasting of.

This is fine weather for shore folk, but not for homeward bound vessels. Lat 5°4' S, Long 18°50' W

Wednesday, March 29th

Steering north-northwest, three sails in sight. Moderate southeast trades during the night better than none at all. At 3 AM passed a vessel supposed to be a slaver. We passed within 20 rods of her and she appeared like a huge white cloud, canvas set upon every available stick on board. This morning served the main-topgallant backstay and seized the fore-topgallant.

Lat 5° S Abominably slow poking.

A rod is 16½ feet, so they were about 300 feet from the vessel they passed in the night. Since they were still south of the equator, if it was a slaving vessel, it was likely en route from Angola to Brazil, which continued to receive shipments into the 1860s and did not fully abolish slavery until 1888. The other possible route would be from the region of the Congo River to Cuba, which continued to receive shipments until 1867 and did not fully abolish slavery until 1886.

Tuesday, March 30th

Rain squalls in abundance. Set up the fore-topgallant backstays. This AM cut off & pudding topped the main shrouds—4 sails in sight. Noon, wind dying away. Steering NNW.

Lat S 3°16' Long W 22°30' double quick.

The foretopgallant backstays run down on either side from the top of the topgallant mast. A pudding was a padded canvas wrapping, in this case of the main shrouds where they passed over the tops.

Wednesday, March 31st

Making spun yarn, the bane of sailors' jobs, putting the finishing touches (pudding tops) on the fore topmast and topgallant stays.

Light breeze, pleasant weather. Steering north-northwest which course with variations of compass, currents, and leeway, makes a north-west course, on a bee line for *New York*. But this does not signify that we are going straight home; it may only be to avoid the Cape de Verd Islands, then we will have to go due north to reach the "Azores" where the captain speaks of cruising, or perhaps in taking a circuitous route home he thinks we may fall in with whales. Let this be as it may—I want to go home and so does the whole ship's crew, and I think there are but few who will exert themselves for more grease.

During the night the wind was quite variable, with some rain for variety in the variations. This morning the sailors are employed in painting the fore, main and mizzen chains; and we are *setting up* the mizzen rigging—one sail in sight, a ship with a spritsail, rather old fashioned, however anything and everything to catch a breath for home.

Lat 7°12' S, Long 23°17' W

By chains, Weir refers again to the thick, narrow platforms on a ship's sides outboard of each mast. They add width to the vessel to widen the spread of the shrouds supporting the masts. A wooden deadeye (a round wooden fitting with three holes) was attached to the foot of each shroud, with a corresponding deadeye at the chains, secured by a long iron chainplate strap to the hull below the chains. A stout rope lanyard was fed through the deadeyes to tighten the shrouds. A spritsail was a square sail set underneath the bowsprit. As Weir remarks, spritsails were largely outmoded by the 1850s as masts grew taller and carried more square sails.

Thursday, April 1st

Steering north-northwest, with light and variable winds. Setting up the mizzen rigging, worked till 5 PM. And after supper scrubbed decks—great amusement? for it makes such pleasant growls arise from the sailors. "The life of the galley slave and a home in a prison ship etc." This forenoon there were seven sails in sight, one of which has been in company with us for several days past, the remainder crossed our course steering west, both ahead and astern of us. By noon steering northwest, the steward caught an enormous Albacora weighing 150 pounds.

9 AM crossed the Equator—very light winds prevail, though what there is, is fair, and we are thankful, for this is the place for calms. Lat 0°13' N, Long 25°17'

An albacora, or albacore tuna, is a large, fast-swimming, predatory fish, with large eyes and long pectoral fins, that swims in schools, often with skipjack, yellowfin, and bluefin tuna. They mostly eat squid, but will also take small fish and crustaceans, often diving deeply to find food. Albacore may grow to more than four feet in length. They were not sought commercially in the United States until 1903.

Wednesday, April 2nd

Working on the mizzen rigging. Caught a porpoise. Light breeze, steering north by west. Now time begins to wear slowly. Oh! for a pleasant gale, to drive our trusty craft back to that sacred land. We can now sing "Home again" and appreciate the words without an exertion—but shall we see that happy land of freedom?

During the night had moderate breeze from the south and east, by 7 AM hauled south—chock aft. This morning rattled down the starboard side of the mizzen shrouds, and set up the mizzen topmast backstays—one sail seen. Lat 1°32' N

Saturday, April 3rd

Steering north by west, with splendid weather moderate breeze, making about 5 knots fixing the mizzen rigging man o'war style. This morning *rattling down* the starboard side of mizzen rigging. Two sails in sight. Begin to get impatient, want to see the North Star—then we shall feel as though we were actually nearing home—home. Lat 3°8' N, Long 26°27' W

Sunday, April 4th

Finished the mizzen rigging and wet hold, after which we were just one hour scrubbing decks—used up two buckets of sand, gave our illustrious mate a chance of illustrating himself. Notice a great many fish about—bonita, Albacoras, dolphins, flying fish etc. Variable winds during the afternoon. At 4 PM steering north-northwest, wind from northeast shifting to the northward. 5 PM braced up sharp heading between northwest and north-northwest. At midnight commenced raining with squalls and cats paws combined; wind quite variable. At 6 AM set the fore topmast

and lower studding sails, wind from east and north steering north by west. At noon moderate breeze from northeast, steering northwest by north. Nothing particular, excepting we did not go the church, as usual with Sabbaths on ship-board.

Lat slow, Long 27°30' W

Sand was used to scour the deck in the process of cleaning it. A bonito is a mackerel-like fish related to tuna. Cat's paws are small, localized patches of ripples on the surface caused by small gusts of wind.

Monday, April 5th

4 PM the weather had cleared off, finely, leaving a fresh northeasterly breeze, by midnight wind shifted to north-northeast or thereabout, still fresh, took in the studding sails and braced up sharp heading northwest.

No sail or land can now be seen to break the even line of the horizon: all around is that same dreary and desolate waste of waters, that we have so long and often beheld, but it is wrong to say waste of waters, for what has God made that is not useful and necessary to the comforts of man, ungrateful creature; it is sacrilege to call it so.

This morning stood look-out at the fore and main mast heads. Ah! Captain your hopes are vain—there is no more oil afloat for us, so rest easy. Lat 5°24' N

Tuesday, April 6th

Head northwest or so with fresh breeze from north-northeast, sailorizing here and there. We are now far enough north to see the Pole Star, but the weather has been so cloudy since we have passed the Lat of 3° N, that to see any star so near the horizon as this would be, is next to impossible. The dipper I have noticed so far south as 24°, and then just over the horizon. 10 PM wind fresh and steady from northeast, steering north-northwest. 6:30 AM steering northwest by north. We move 8 knots full—this looks like going home. Hands employed in making spun yarn from tarred towline yarns, anything to keep them busy. From 6 to 7 AM I brightened and sharpened my harpoons and lances in hopes of not being obliged to use them again. Fine breeze continues and all feel jolly. Lat 6°34' N, Long 31°17' W

Home again.

Wednesday, April 7th

Manufacturing spun yarn and the boys enjoy it amazingly? Seized the main stays and gave the decks an awful scrubbing, to satisfy an abominable humor of the mate,—he drilled us all for an hour and a half—what a delectable situation for person of refined feelings to occupy, such boorish and uncivil treatment is truly gratifying? for there is so much to be gained from it. Blow winds and drive us home from this. 6 PM took in the studding sails and braced the yards, steering northwest by north, wind from northeast. Had a right stiff breeze most of the night. By daylight steering northwest, broke out molasses for the sailors.

Lat 8°15' N, Long 33°37' W

In describing the arduous duties of the day, Weir expressed his sarcasm with question marks after the positive words.

Thursday, April 8th

Steering northwest with a smacking topgallant breeze. At dinner the captain presented me with a bunch of segars, which were truly acceptable for I have been out of pipes, tobacco and segars since leaving the Indian Ocean. Nothing particular doing. Of late I have noticed great quantities of Portuguese men o' war about. During the night the sailors caught several flying fish, but I was not fortunate enough to get any: it is a great while since any have boarded us, but I shall loose out for a few wings to carry home with me, for though these fish are as plentiful here as birds on land they are supposed by many to be a fictitious creation. The largest

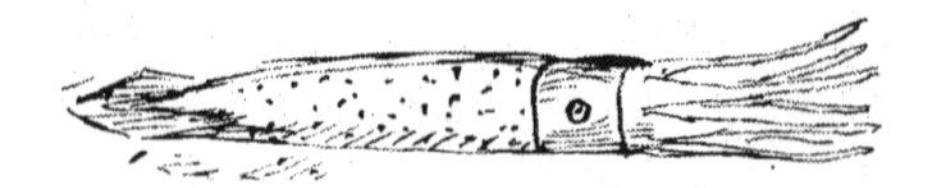

flying fish we have yet seen did not exceed 14 inches. I have heard of their being larger than that, but am inclined to doubt the veracity of such statements. The smallest could scarcely span ⅞th of an inch. And how wonderfully they are formed, the mightiest whale is not more so.

The most curious fish is the cuttle or squid as the whalemen call it, the flying squid particularly, for they propel themselves through the air by suction and expansion in a truly strange manner. Had a smacking breeze all night. 6:30 AM set the fore-topmast studding sail wind from southeast by east. Being little to do, set the hands at work pecking the rust from the anchors and windlass bars. 6 AM steering northwest by west. Made new fore spencer vangs.

Lat 9°35' N, Long 30°36' W

The Portuguese man o' war is a jellyfish-like siphonophore with a pneumatophore—large, inflated membrane—that floats on the surface and acts as a sail. Tentacles below have stinging, venomous nematocysts. Cuttlefish and squid are two separate species of multiarmed cephalopods that move by forcing water through the mantle. Squid can jet at up to 25 miles per hour. About 18 varieties are known as "flying squid," using their jetting power to break the surface and glide for up to 160 feet to escape predators. The windlass bars are the long iron arms with handles raised and lowered in opposition by the crew to operate the pump-break mechanism of the windlass. A spencer is a fore-and-aft sail with a gaff at its head but no boom at its foot. A whaleship often carried a spencer on the mainmast and sometimes the foremast. The Spencer vangs were controlling lines, running from the outer end of the gaff to the rail at either side.

Friday, April 9th

Steering northwest by west with good breeze. Set up the new vangs and served the bowsprit man ropes. Overhauling paint buckets brushes etc.

etc. etc. Had a fine topgallant breeze all night, blow winds and waft us home! Splendid weather. Had a long gaze at the North Star last night, the sky being clear in that quarter; and what vivid recollection of by gone days it brought before me; how often have I looked upon that star while crossing the noble Hudson between West Point and Cold Spring; or while any where about home and at night. Can I look at that star from that loved place again. I trust it may be so, but I can't anticipate. 8 AM set foretopmast and lower studding sails, broke out water and pork. 10 AM set the weather main-topgallant studding sail.

Saturday, April 10th

Lat 10°42' N, Long 37°40' W

Steering northwest. Sailorizing on a small scale. We are inclined to think the captain is anxiously waiting for calm weather in order to tar down the rigging and paint ship, but I trust we'll get out of these doldrum latitudes first, so far we have done well, in having some kind of breeze ever since leaving port.

Bowling along at a good jog all night, this morning I *set up* a harpoon, an undefinable whim of the mate.

7 AM, steering northwest by west, good breeze continues. Nothing particular going on.

Lat 11°56' N, Long 42°04' W

Sunday, April 11th

Fine weather, moderate breeze, wet hold as usual, and I hope it is the last aquatic evolution that I shall have to perform. Mr. Barker remarked that it would be, and I am perfectly agreeable. Had splendid royal breeze all night, northwest. I'm out of soap, and last night I was obliged to get my clothes washed in clear water, for even the ley-barrel is empty, so in lieu of soap elbow-grease was plentifully applied—thank goodness there will be but few soiled clothes from this present time till, *then*—home, home.

A magnificent day is this—how pleasant to enjoy a walk with Em to the little church. What pleasant rambles we used to have together Em—such joyous times seem not for us again. Though old time may yet go lightly. Lat 13°14' N, Long 44°20' W

"Ley-barrel" suggests that for cleaning purposes the *Clara Bell* carried a barrel of caustic lye, which was then made by soaking wood ashes in water. Lye would be used for cleaning the paint, but it was also the basic component of the soap used on board.

Monday, April 12th

Moderate breeze, steering northwest. The water here has a very light blue appearance, which by Maurys theory denotes the extreme depth. Sounding are marked on the charts hereabouts from 3 to 4000 feet deep. Could we see the waters cleared at such a depth—what awful and mysterious wonders would we behold.

I wonder if the good folk at home have thought of me this day? Had a good breeze all night with royals set. This morning broke out the run cleaning up and irons, lances etc. etc. etc. Painted all the iron-work that belongs to the run.

Lat 14°39' N, Long 46°18' W

Weir seems to refer to the theory that the depth and salt content of sea water changes its perceived color, with saltier water appearing darker blue. This was articulated by the pioneering US Navy oceanographer, Lieutenant Matthew Fontaine Maury, who compiled wind and current charts of the oceans in the 1840s and 1850s and, in 1855, published *The Physical Geography of the Sea.* He discussed water color in chapter one, describing the Gulf Stream.

Tuesday, April 13th

Steering northwest with light easterly breeze. Served and set up the main swifter starboard side—stowed all the iron work in the run between 4 and 5 PM and washed decks after tea. During the night had all sail set with pretty good breeze, this morning broke out groceries, water etc. Tarred down the fore and main royal and topgallant rigging, getting ready for home, now look out for paint and paint pots ye tars. The breeze seems to be gradually leaving us. Lat 15°47' N, Long 48°7'

A swifter is the aftermost shroud, supporting the uppermost portion of a mast.

Wednesday, April 14th

Tarred the fore-topmast and lower fore rigging, breeze freshening from the northeast, 5 PM steering northwest a half west. Excellent breeze all night with smooth sea, and the sails keep a *rap* full all the while, this morning tarred the fore and aft stays. We make 9 knots with this breeze, and it is right cheerful.

Lat 17°30' N, Long W 50°42' W

"A rap full" means the sails are full with a steady wind, often steering on the wind—close-hauled—with the yards braced and all sails pulling well.

Thursday, April 15th

Steering northwest half west with fine breeze from the northeast which has continued all night. Our good bark glides along like a thing of life, and all hands feel influenced by her progress. While at the masthead this evening I noticed a great deal of sea weed about. This serves to show the advance we are making towards the outer currents of the Gulf Stream and home. Oh! my, smoked my last segar during the midnight watch, felt wretchedly homesick—though why should I, not knowing whether I have a home—however I did surely have one once, and suppose I can only brood over the thoughts of it hereafter. But why do I get so low spirited—simply because I can't help it—*avaunt.*

Sailors doing petty jobs here there and everywhere.

Lat 18°54' N, Long 55°58' W

Friday, April 16th

Steer northwest a half west, fresh northeasterly breeze. Sailorising miscellaneously, tarred the bowsprit gear etc. etc.

During the first half of the night the breeze was light, but not so much so as to raise a growl from the fo'castle, during the morning watch the breeze freshened—see plenty of gulf weed

Lat 30°7'N, Long 55°15'W

Gulf weed is more commonly called *Sargassum*, a form of brown microalgae, several species of which collect around the Sargasso Sea, an Atlantic Ocean gyre bounded on the west side by the Gulf Stream.

Saturday, April 17th

Steering northwest a half west direct for home, moderate breeze, all sail set. During the night had fair breeze from east-northeast, kept every stitch of canvas set all night. Begin to feel nervous, or I guess it is home-sickness. Blow! Good breeze, blow! Lat 21°22' N, Long 57°22' W

The *Clara Bell*'s compass had 128 directional points, with the 32 standard named points subdivided by quarters and halves. Northwest a half west is the halfway point between northwest and northwest by west, or approximately 310°.

Sunday, April 18th

Moderate breeze steer northwest a half west as usual. Gave out a few clothes to be washed and for the last time I trust, and Manuel says he will not wash another piece so long as he is on board this ship unless we are becalmed on the edge of the Gulf Stream for a couple of months. Wet hold.

Good breeze continues, at sunset set the royals. About midnight the wind hauled southeast and east, chock aft, nautically speaking. Dreaming of home by night & thinking and talking of home by day, and so it is with all the crew. But will they, can they ever love me as formerly? If not, no one is to blame but my most willful self—deeds, not words must hereafter mark my life—how shall I face my honored and good father, to obtain his forgiveness is all I shall seek, and my future life must show my gratitude.

Lat 23°58' N, Long 59°27' W

Monday, April 19th

Light winds from the eastward, how I should like to walk to church this afternoon with Em. I wonder if they will think of me this day, well! there is some satisfaction in imagining they will—flattering thought. At 4 PM breeze freshening, steering northwest. Put the *harness* cask on deck this morning, washed and scraped it for painting, mixed paints and got things ready to start after dinner. I notice the seaweed is not so plentiful as formerly. Lat 24°10', Long 61°9'

The harness cask was a large cask in which the salt beef and salt pork were placed before cooking and after being broken out of the storage casks in the hold. The harness cask was normally kept on deck near the galley, but it may have been stowed below on the *Clara Bell* so as not to impede whaling operations on deck.

Tuesday, April 20th

Steering northwest. Took in the staysails and studding sails in order not to incommode the painters. Commenced painting or daubing 'i the eyes and head, about the windlass forecastle and etc. 5:15 PM finished for the night, had supper and took in the flying jib and gaff top sail, also the mizzen topmast staysail and fore topgallant sail. Wind shifted north-northeast, weather looks threatening, of course, on account of the fresh paint. We managed to head northwest all night with yards braced up sharp.

At half past five this morning, good weather, commenced painting—all the watch going at it, hammer and tongs, till 7:30 AM when the Larbonians relieved us. Had quite an interesting talk with Dewey last night; his situation is so like my own and his former harrum scarrum life agrees so much with mine, that we find considerable pleasure in making plans, to live a new life hereafter and to honor our parents. Oh! what sorrow I know I have caused my noble father. I trust I shall hereafter know my duty better, and by Jove I shall.

Lat 25°20' N, Long 62°43' W

The "eyes and head" were the very forwardmost parts of the hull. Traditionally sailors used this area as their toilet; thus, the term "head" for seagoing toilet room. Whaleships were beginning to carry a toilet stall on deck near the bow but there is no evidence that the *Clara Bell* had one. "Dewey" was Charles P. Dewey of Rochester, New York, who sailed as an 18-year-old greenhand on the bark *John Dawson* in 1855. He left the *John Dawson* and joined the *Clara Bell*'s crew at St. Augustine Bay, Madagascar, on February 8, 1858. He kept a journal that contains two pencil sketches by Weir. (Log 590a, New Bedford Whaling Museum, New Bedford, MA.)

Wednesday, April 21st

Still steering northwest, have not made any more sail yet, three sails in sight; 3 PM a schooner running across our bows—we hauled aback, and lowered a boat for news, proved to be the *Charm* from Philadelphia, ten days out, they tossed us a package of papers, which proved a grand treat and serves to make us realize that we are actually drawing near home. This schooner is bound for Santa Cruz—she sails like a pilot boat, and any way she is not much larger. By evening we had finished painting the bulwarks. This morning two sails in sight; set our colors for an Englishman, but ran too wide to speak.

10:30 AM wind fair from the southeast, steer northwest. Set fore topgallant sail, flying jib and gaff top sail, painting the bearers, davies, etc., etc. Notice plenty of Gulf weed about.

Lat N 26°22' Long W 64°11'

This may be the 73-ton coasting schooner *Charm* from the Delaware River, commanded by Captain Blizzard. Though a small vessel for an ocean voyage, it sounds as if the *Charm* was headed for Santa Cruz, Tenerife, in the Canary Islands off Africa. A swift-sailing vessel like this may have been engaged in the fruit trade. The bearers are the vertical posts along the rail between the davits to which are attached the hinged cranes that support the whaleboats. "Davies" means the whaleboat davits at the rails. The *Clara Bell* had four pairs of davits and four pairs of bearers.

Thursday, April 22nd

Steering northwest with light southeastly breeze, under topgallant sails. Oh! dear! How can the Old Man keep us reduced to this short sail with so light a wind not more than four and a half or five knots. 5 PM finished painting, set the studding sails alow and aloft—hurrah. Very light breeze this morning, can't knock more than three knots out of our craft, however that is preferable to a dead calm. Steering north-northwest, plenty of weed about yet.

Lat 27°9', Long W 64°37'

Friday, April 23rd

Almost dead calm, steering north-northwest. 5:30 PM, a light breeze sprung up from eastward, set every rag, painted the topsail sheets, scraped and varnished the wheel etc. Also painted the starboard side of the ship. Moderate breeze—beautiful weather.

Lat 27°50' N, Long 65°10' W

Saturday, April 24th

Steering north-northwest, with a villainous three knot breeze, sailors muttering and growling. Painted the larboard side of the ship. Oh! blow! blow ye winds and waft us home. Touched up scientifically here and there on the inside of the bulwarks.

Wind increased during the night, by day light fair breeze from the southward. Studding sails set alow and aloft. Scraped and varnished the lash-rail and spanker boom, scraped all the masts—*Clara* begins to look new indeed. Brewster, the owner, will congratulate himself on having left his petted craft in the hands of such a skipper.

Lat 29°46' N, Long 66°7' W

Sunday, April 25th

Quite squally, steering northwest by north, took in the light sails to topsails. Bent, rove, hauled taught, and belayed the topsail sheets. 2 PM set the light sails again. 8:30 PM squally and unpleasant, took in sail to topsails, raining 60 guns. Felt quite unwell last night, and at 7 PM took a dose of some villainous mixture and *turned in* to my berth, it being my watch below, at about a quarter before 9 I was awakened by hearing the greatest tramping about decks, the rattling of ropes and above all, the captain's voice giving orders to take in sail—the wind had hauled to the north-northwest bringing somewhat of a severe squall so they had to take in all light sail and luff by the wind on the larboard tack, heading northeast—the rain fell in torrents, by daylight set all sail with fair southeast breeze. Lat 30°50', Long 67°15'

Monday, April 26th

Steering north-northwest, with fresh southeasterly breeze, the weather is unsettled, and between 3 and 4 PM caught a short head sea which has set us galloping in an uncomfortable manner. By sunset, moderate breeze, sail in sight ahead, and we are overhauling her fast. The wind seems to be playing hide and seek around the sou'western corner of the compass, for it has been coming from any point between south and north-northwest most of the afternoon. 10 PM, breeze light but freshening—heading northeast.

2:30 AM tacked ship heading west by south. We are now close hauled to the wind with fine topgallant breeze, the sea is just rough enough to keep one lively—at least a sailor. Watch employed in painting the boats.

Lat 31°36' N, Long 68°20' W

Tuesday, April 27th

Breeze fresh from the eastward. steering north-northwest. By 8 PM had fine southerly breeze. During the midnight watch we managed to make five knots with every rag set, studding sails and royals just rap full. 8 AM—hurrah! we have a smacking breeze, and it is still on the increase as it has been all night. Made a new spanker-out-haul-block for the heel of the spanker boom; the last job I hope to do this voyage.

Very little gulf weed about now.

At noon took in the studding and topgallant sails, though it don't blow more than a good topgallant breeze. Watch employed in cleaning the deck; scraping the paint off etc.

Wednesday, April 28th

Steering north-northwest under full topsails, when we could easily have topgallant sails and topmast studding sails set, the sea is getting rugged. 2 PM blowing great guns, topsails double-reefed, mainsail, jib, and spanker furled. Wind hauling to the westward. 4 PM, blowing a perfect gale, furled the fore-topsail, close-reefed the main-topsail, and reefed the fore-sail, and a hard time we had doing it too: one of the topmast studding sails booms got loose and came nigh sweeping half a dozen of us overboard. Ship heads anywhere between northwest and north.

7:30 AM been blowing a heavy nor'wester all night—though the thermometer notes 60 degrees; we suffer with the cold—what shall we do in wintertime?

A sail passed us about midnight scudding under full topsails—the wind had now knocked us off north-northeast. At daylight discovered a sail not above a mile off our lee beam, she is a brig, with a reefed mainsail, reefed foresail, double-reefed fore topsail and main staysail—she forges ahead of us, but drifts badly.

Had eight hours on deck last night, and a cold comfortless watch it was, with some rain and plenty of salt spray.

7 AM set the fore topsail reefed but had to take it in again.

Saw a lunar rainbow during the latter part of the first watch, as the moon showed her face quite often. And I could not help thinking that some of the dear ones at home were enjoying her soft light—if not, there is some satisfaction in knowing that this same moon was shining in many familiar spots as I gazed upon her. By noon we head our course north-northwest under double reefed topsails, jib, spanker and mainsail. Weather still rugged and unsettled. Lat 34°16' N, Long 69° 15' W

Thursday, April 29th

Good! it *looks like clearing*. 2 PM set jib, spanker and mainsail, 4:30 PM shook the reefs out and set the main topgallant sail. Two sails in sight.

After 6 PM quite squally, had to take in the main topgallant sail and keep our weather eye open. 7 o'clock blowing a gale, increased every moment, 7:30 PM blew sixty guns; clewed up the fore topsail, clewed down the main topsail, hauled up the mainsail and took in the jib and spanker, and hauled up the foresail. Fore-topmast staysail tack parted, and I was sent out on the jibboom to lash it down with an earring—at this time it blew so strong that it was impossible to look to windward—yet, all hands worked cheerily for they know home is but a short week off. 8:30 PM, the gale had blown over somewhat, set the foresail—it was an unpleasant night—raining most of the time. All hands talking about that place of places—home sweet home. Head anywhere between north and north-northeast, not very satisfactory for home is not ahead but astern.

The weather is cold, thermometer 56 degrees, and everybody seems to suffer. The water in the butt tastes as though it were well iced, for it makes one's jaw ache. 7 AM, set jib, spanker and mainsail steering northwest—home right ahead.

Lat 35°3' N, Long 68°54' W

The fore-topmast staysail was the triangular staysail or jib set on the stay running from the head of the fore-topmast to the jibboom on the bowsprit. The tack was the lower corner, and the earring was a line to secure the tack to the jibboom so the sail could be set again. Weir does not mention the risks in venturing out on the bowsprit in such conditions.

Friday, April 30th

Steering north-northwest, fresh breeze from the westward. Whew, cold, thermometer tells 50 degrees. A clear cold sky, haven't felt such cold weather for three years.

All hands overhauling the round house. By 4 PM set all sail with the exception of studding sails, but had to reduce it almost immediately to double reefed topsails, scrubbed decks with sand, water awfully cold, and we feel it so much after having been in warm latitudes so long. Strong wind from northwest. 7 PM wind variable—braced round the yards, heading southwest by west. Splendid moonlight night. 2 AM tacked ship heading north. By 5 AM had all sail set steering northwest by north. This morning tried the temperature of the water, thinking we might be in the Gulf Stream, water 1 degree warmer than the air. No Gulf Stream yet, but shortly will be.

The steward played a nice trick on several of the boys—he got some boiling water from the galley and poured it into the ships bucket, then held it over the side as if he had just drawn it and exclaimed "Hurrah boys we're in the Gulf Stream now—just feel this water!" Several green ones rushed to the side, and almost simultaneously thrust their hands into the bucket, but drew them back as quick, but not without having been seen by a number of the sailors, there was an uproarious laugh just then which they did not seem to enjoy.

Lat 36°20', Long 69°30'

Saturday, May 1st

Steering northwest by north a half north, smacking breeze from west-southwest. We glide along splendidly, and all hands wear bright faces. Before 2 PM took in the studding sails, unbent and sent down the main top mast and topgallant staysails with the preventer stays. 6 PM, shortened sail to topsails. 7 PM double-reefed the foretopsail. 12 midnight we are beating about in a very rough and irregular sea—the water is of a dark slate color. The skipper tried the water several times this afternoon, but with no satisfaction, if our reckoning be correct we are now on the very edge of the Gulf Stream.

8 PM, double reefed the main topsail, but not without some growling.

Stood under full topsails for about two hours during the middle watch. Between 1 and 2 AM I noticed a small double ring about the moon—it seemed to me to represent the hues of the rainbow, though there were but two others who thought they could distinguish colors. Captain dipped the thermometer quite often during the night. Daylight made all sail heading northeast or thereaway. 10 AM, three sails in sight, one of which is a whaler and the captain says it is the *Almira* though the mate doubts. 10:30 AM tacked ship heading west-northwest. A good course if we are actually in the stream. Splendid weather—clear, cold and bracing—but all are so fraught with anxiety at this present time that they can't enjoy such blessings—all are thinking and dreaming and talking of home.

Lat 37°30' N, Long 69°19' W

Sunday, May 2nd

Wind from southwest. Excellent weather, head west-northwest, fresh breeze but not fair enough. By 4 PM steering northwest by north, *Almira?* off starboard quarter. 5 PM set the royals—weather looks bad—we are now in the centre of the Gulf stream, or pretty nearly so. 7:30 PM set fore-topmast and lower studding sails. Steering northwest, breeze freshening. Our passage through the waters is like a comet—there is such a brilliant train of phosphorescent light behind us, it contrasts brilliantly with the surrounding darkness, this illumination does not appear like the general lights of the waters; it has a ghastly look, a bluish white, while elsewhere there is a warm pink or yellow tinge.

Between 10 PM and 1 AM, the water around the ship was splendid to look at, it flashed and sparkled far beneath the surface, looking like so many twinkling stars in the heavens.

About two hours before the moon rose, we had a sight which brought many home recollections to our minds. It was the Aurora Borealis. Everything now indicates our close proximity to our native land, to our glorious home. We see the North Star almost as high in the heavens as they do at home, and there is not more than six or eight minutes variation in the time compared with home—hurrah—we are blessed indeed. During the midnight watch set the starboard main topgallant studding sail, steering northwest by north, had a smacking royal breeze all night. By daylight an unknown whaler off our weather bow.

7:30 AM steering northwest by north. A stormy Sabbath—11 AM, had two reefs in the topsails, horrible weather—hail, rain, and cold winds. At noon bore down and spoke the ship *Huntsville* of Cold Spring, Long Island, compared chronometers—nearly eight minutes difference. The skipper thinks we are in the wrong. Too cloudy to take an observation.

The ship *Huntsville* of Cold Spring Harbor, New York, commanded by Captain Grant, departed for the North Pacific whaling grounds on October 15, 1854, and returned on May 6, 1858, with 651 barrels of right or bowhead oil, having sent home 1,457 barrels of oil and 32,035 pounds of whalebone (baleen), probably from Hawaii.

Monday, May 3rd

2 PM weather is growing better, 4 PM set the topgallant sails, 6:30 weather quite fair, moderate breeze from north and east, steering northwest.

Hauled aback the head yards and hove the deep-sea lead, found 60 fathoms, 11 PM, 49 fathoms, 1 AM, 45 fathoms, very light wind, and very cold weather, 3 AM wind increasing, steering north by east. Took several observations from the stars, had somewhat of a dispute with the mate in regard to the North Star. By these stella observations we are too far westward of our course. 4 AM tacked ship heading east by south. 6:30 AM tacked heading north and north-northwest. This morning took all the whaling craft from the boats and stowed it below, stripped all the chafing gear from the rigging etc. Two schooners in sight—but where is the land. We suffer with the cold, though the thermometer stands 50 degrees, they'll laugh at us at home—but if we can actually get there, let them laugh and will join in with a hearty good will. Lat 40°25'

A lead line was a long line with marks at fathom (six-foot) intervals and a heavy lead weight on the end. The weight had a cupped bottom into which tallow could be placed to bring up a sample of the bottom material. It would be cast forward and allowed to run out to touch bottom, indicating the depth. In shallow water, a line of about 25 fathoms, or 150 feet, was used. The deep-sea lead was used offshore and was about 100 fathoms (600 feet) long. Knowing the depth and bottom material, Captain Robbins could consult the chart to approximate their position on the continental shelf. It appears that Weir had learned to use a quadrant or sextant for celestial navigation and to perform the calculations, leading to his "dispute" with First Mate Barker.

Tuesday, May 4th

Steering northeast, moderate breeze, fine weather, clear and cold, set the main royal at 1 PM. 2 PM set the fore royal, 3 PM set the fore topmast and lower studding sails. 4 PM set the main topgallant studding sails. Breeze dying away—what shall we do? The land ought to be in sight by our reckoning, curious navigation this is. The Old Man seems to be perfectly indifferent as regards getting home—he must be case hardened, or else he is putting these contrary actions upon himself to tantalize us.

5:30 PM, joyful news the reviving cry of Land ho! has been uttered, and three hearty cheers made the welkin ring again. My pulse beats warmer, and though I have eight hours to watch on deck this night, it will be done cheerfully, for I feel it to be the last weary watch for me to hold. The wind is quite light but fair, and increasing somewhat. 7:30 PM dark, raised Montauk light, 8 PM took soundings, made 35 fathoms, tacked ship and kept lookouts at the masthead all night, in order to keep the light in sight. I had one dreary cold hour to spend there myself with the second mate continually disturbing me while thinking of home—he talked about his wife and child, but I couldn't listen to him, for I had my own thoughts for better company. Took in studding sails and topgallant sails and tacked several times before morning. At midnight the ship *Huntsville* was close to, they sent a boat alongside for a short gam and comparison of reckoning. The weather is villainously cold, enough so to keep us all in an uncomfortable state. 4 AM day break, made all sail. 6 AM six sails in sight, mostly schooners, 6:30 pilot boat *Relief* bearing down to us, hove to with main yard aback and received the pilot, braced forward and steered northeast. Wind light from west, several islands in sight—Martha's Vineyard, Gayhead right ahead, Rhode Island off larboard beam.

7 PM we touched the wharf and I touched the shore—pure bonafide American land hurrah.

We are now safely moored at one of the New Bedford wharves—it is glorious to think of. But I am thankful, we had a favorable breeze to come in with, and the pilot took us chock into the wharf.

9 PM I have just returned from a stroll on solid Yankee land.

How shall I face my dear father—I shall go directly to him and tell him all. I trust God will yet give me strength of resolution to reform.

Montauk lighthouse is at the eastern end of Long Island. Second Mate Cyrus Clark, formerly a master and then first mate of the *Brewster*, must have had a strong will to get home to his wife and child in order to switch to the *Clara Bell* as second mate. The *Huntsville* would steer north from Montauk and then west up Long Island Sound to reach Cold Spring Harbor, Long Island. The *Clara Bell* continued east until they sighted Gayhead (now Aquinnah) at the western end of Martha's Vineyard, about 25 miles south of New Bedford. Each port had its pilots, who were skilled in navigating the winds, currents, tides, rocks, and shoals that complicated entry into the harbor. New Bedford's pilot schooners, including the *Relief*, cruised off the coast between Block Island, Rhode Island, and Nantucket Shoals to intercept whaleships. With a pilot on board, the whaleship could then safely approach and enter New Bedford Harbor. The pilot worked the winds and tide to sail the *Clara Bell* right to the wharf, without use of a towboat.

~End of Journal~

came up suddenly
hit him in the small

Poems

May 18th, 1857

Trust not thyself

"Resist in time; all medicine is but play
When the disease is strengthened by delay"

On different senses different objects strike:
Hence different passions more of less influence
As strong or weak the organs of the frame
And hence one masterpassion in the breast
Like Aaron's serpent, swallows up the rest.

For every trifle scorn to take offence
That always shows great pride, or little sense,
Good nature and good sense must always join
To err is human—to forgive divine. Pope

Weir quotes fragments of Alexander Pope's acclaimed 1733 poem, "An Essay on Man."

Trust not thyself but thy defects to know;
Make use of every friend, of every foe. Rev. (??? R.W.Walker?)

Glide in thou brilliant orb of night
To show thy bright face home
Tell them of those who are in order here
Where friends can seldom come
The far from friendly shore

Carry our thoughts upon thy beams
And all our silent prayers
All wishes we have for their good
But do not show our care

And with they light take this request
That they will pray for us
Tis needful for our hope of rest
Lies only with our God

These thoughts take home with thy bright face
And scatter with thy beams
Among all friends that we have left
Who see us in their dreams

From his own fair childhoods home
And his gentle Annie's eyes:
He rests within the blue below
'Neath the blue of foreign skies
The dashing wave is his last home
And the spray his winding sheet,
The coral is his headstone high
And pearl is the stone at his feet.

He hears no more the wild watch cry
Nor shouts the sail ahoy
The quick life's pulse in his breast is still
And it throbs not with pain or joy
The snow falls not on his silent couch
But the wind whistles cheerily over
The tomb that is fit for a seaman's child,
The resting place of the rover.

He loved the sea in his dreamy life,
And he could not rest in peace;
If all the sounds of ocean strife
And the waves wild march should cease.
He could not rest with the wildflowers fair
And violets o're him bending
With gentle breath of the evening breeze
Their perfume sweet were lending.

Up would he start from his restless sleep
And up from his cold earth's pillow.
And hie him away to the dashing sea.
To a grave beneath the billow.
There shall he rest till the last dread day
And the summons that comes to all.
Then e'en from the sea the dead are piped
Aloft by the last trumps call.

This has not been identified as a published poem so it may be Weir's own attempt at a romantic poem of a lonely boy at sea. The phrase "his gentle Annie's eyes" may refer to his future wife, Anna Chadwick.

[The Pilot]
Oh Pilot! tis a fearful night
There's danger on the deep;
I'll come and pace the deck with thee
I do not dare to sleep
Go down! The sailor cried, go down –
This is no place for thee;
Fear not but trust in Providence
Wherever thou might be.

On such a night the sea engulphed
My father's fearless form,
My only brother's boat sent down
In just so wild a storm.
And such, perhaps – may be my fate.
But still I say to thee;
Fear not! But trust in Providence
Wherever thou may it be

Weir quotes the English song "The Pilot," written by Thomas Haynes Bayly and composed by Sidney Nelson in 1831. In the second line of the last verse, Weir changes Bayly's wording—"my father's lifeless form"—to "my father's fearless form."

It is more beneficial to live in subjection than in authority
And to obey is truly much safer than to command.
Why should I murmur now that I am in subjection
Rash boy be resigned it is the will of God
And this I know is for my good.

This has not been identified, so it may be Weir's musing.

Postscript

The *Clara Bell* returned 971 barrels (30,586½ gallons) of sperm whale oil, 260 barrels (8,190 gallons) of right whale oil, and 1,900 pounds of "whalebone," the industry term for the baleen from a right whale's mouth. At the average prices for oil and baleen at that time, the cargo was worth about $45,030 (about $1,684,600 in 2023). Weir's share would have been only about $325 (about $12,160 in 2023), from which were deducted the cash and sailor's outfit advanced to him at the outset, plus the cash and personal items (slops) he drew during the voyage, and interest on all these advances. As a result, he probably received about half of what he had earned.[1]

Nevertheless, Weir was proud to have drawn only $80.53 (about $3,000 in 2023) in cash and slops from the *Clara Bell* during his 33 months at sea. He itemized his expenses: almost 30 pounds of tobacco, two pounds of it "miserable"; a sheath knife and three jack knives; a sailmaker's palm for mending sails; "shaving apparatus"; nine pairs of shoes, including a "confounded poor" pair, an "awful" pair, and a pair of light sailor's pumps; 10 pairs of pants, including two pairs of heavy duck pants; eight woolen shirts and two cotton ones; two straw hats and one sou'wester for foul-weather wear; a coat and an overcoat; about 20 pounds of soap; six yards of "Madagascar Calico," or trade cloth for barter at Madagascar; and $2.00 "wasted at St. Helena," $5.50 spending money at Mahé, and $4.00 cash at Johanna. Despite his frugality, he wrote: "I was happy and contented all that time, as much as man can be when thrown upon the world."[2]

Soon after the return of the *Clara Bell*, Captain Robbins would take his wife Hannah to sea with him on the whaling bark *Thomas Pope*. Their daughter, Lizzie Pope Robbins, was born during the voyage in 1862. He would then command the *Kathleen*, the *Cape Horn Pigeon*, and, finally, the *General Scott* for its 1875–1878 voyage. While Robbins was at sea during that last voyage, his old command, the *Clara Bell*, was trapped by ice and abandoned near Point Barrow, Alaska, in the western Arctic in the summer of 1876. When the ice broke up in 1877, the vessel drifted off and was lost.

After retiring from whaling, Captain Robbins became a handyman and housekeeper for a New Bedford police station. He also began writing down his reminiscences and whaling stories. With the assistance of his daughter Lizzie, they were finally published in 1899 as *The Gam, Being a Group of Whaling Stories*, which included incidents from the *Clara Bell* voyage. Charles H. Robbins would die at New Bedford in 1903.[3]

Robert F. Weir left no record of a reunion with his family, and his whereabouts from May 1858 to April 1861 are unknown. In October 1858, a Robert Weir sailed on the whaleship *Lancaster*, giving his age as 23 and home as Hartford, Connecticut. The ship whaled in the Pacific and, during its return passage, was condemned at St. Thomas in the West Indies in 1861. It is more certain that Robert F. Weir was, in fact, the Robert Wallace who joined the whaling schooner *Palmyra* as a boatsteerer in April 1861.[4]

The 22-year-old, 100-ton schooner *Palmyra* was a very small Atlantic whaler. The master, 39-year-old Edward S. Davoll of Westport, Massachusetts, was an experienced whaleman who had gone to sea in 1840 and took command of the bark *Cornelia* in 1848. In 1855, he lost the bark *Iris* in a storm on the west coast of Australia. Arriving home in 1856, he then took command of the *R. L. Barstow* and gammed with the *Clara Bell* near St. Helena in March 1858.

In 1860, Captain Davoll became involved in the effort to outfit the large whaleship *Brutus* to engage in the African slave trade. American participation in the trade in enslaved Africans had been prohibited since 1800, but in the late 1850s, a number of New York and even New England merchants and captains decided it would be a lucrative, illegal use for their underutilized vessels to deliver Africans to Cuba (and even to the American South). Davoll helped outfit the *Brutus* ostensibly for whaling and sailed in command but then left the vessel in the Azores, after which the mate took the *Brutus* to the Congo River and loaded 600 Africans. When they arrived at Cuba, only 500 were alive. After news of the voyage reached New England and prosecutors investigated, one of the outfitters was indicted, but Captain Davoll and the other outfitter were not charged. To confirm their legitimacy, they quickly fitted out the *Palmyra* for an actual whaling voyage.[5]

As a two-boat whaler, the *Palmyra* carried a small crew:

E. S. Devoll, master
Charles B. Shaw, first mate
David Mindell, second mate
Prentiss M. Bearse
William Robinson, boatsteerer
Robert Wallace, boatsteerer
Robert Williams, cook
Antone Jose Alvias
George Eaton
Thomas Gardner
John L. Chase
John Jackson
John Jay
Salomon Jones
James Kelley
Manuel Levias[6]

The *Palmyra* departed New Bedford on April 24, 1861, 11 days after the surrender of Fort Sumter off Charleston, South Carolina, signaled the beginning of outright conflict between the Union and the new Confederate States of America. The vessel first headed for the Azores to obtain further provisions and to seek sperm whales among the islands. The logbook recorded Captain Davoll's admonition to the crew: "You who want money will please bear in mind that a standing Bounty of ten dollars for every one hundred barrels of Sp [sperm] oil stowed down on the voyage will be rewarded to the man who first Sings out for the whales & reports the same to the deck." By August, the *Palmyra* had left the islands and was cruising in the Atlantic, where the whaling was poor.[7]

As he had earlier written, Captain Davoll especially enjoyed "smoking & chewing tobacco, telling stories with the mates and Boatsteerers."

However, on September 4, Davoll came on deck at 3:00 AM to find Boatsteerer William Robinson asleep during his watch. Davoll grabbed him and ordered him into the cabin, calling the mates to help secure him as he fought back. They brought him on deck and seized him to the shrouds, ready to receive a flogging. The US Navy had outlawed flogging in 1850, but some whaling captains still used it to enforce discipline. Davoll administered three lashes before Robinson "begged like a dog" and was released after promising to do his best thereafter. It is not known why Davoll resorted to the most serious punishment for a relatively minor infraction among a small crew, an action that would only inflame dissent.[8]

The next day, the whaleboats took three whales, producing 87 barrels of oil, and on October 17, they sighted a 100-barrel whale. One boat was able to harpoon the whale and fire a bomb lance into it, but it pulled the boat off on a Nantucket sleigh ride, and after an hour, the mate cut the line. "Went to the Devil," recorded Mate Shaw in the logbook. As the weather then worsened and the seas rose, smashing one of the boats on the davits and filling the deck with water, Davoll headed the schooner for the neutral British island of Bermuda, where they would be safe from Confederate commerce raiders. As they neared the island on November 7, 1861, the other whaleboat was stove when one of the supporting cranes gave way.[9]

Davoll planned to keep the *Palmyra* at Bermuda for the winter, possibly in fear that if he returned to New Bedford he might face trial for his role in the slaver *Brutus* in 1860. He transferred the *Palmyra*'s 35 casks of oil to the neutral British schooner *Princess Royal* for delivery to New York. Within a few days, Second Mate Mendall refused to perform any more duty on board and moved aboard the British schooner. Boatsteerer Robinson and cook Williams also left the *Palmyra*. It appears that the rest of the crew then dispersed. Weir may have found his way back to New York on the *Princess Royal*.[10]

The Civil War was entering its second year when Robert F. Weir returned to the United States. His younger brother Henry was a cavalry staff officer, and their brother Gulian was an officer in the 5th US Artillery, while their brothers-in-law also served: Truman Seymour was promoted to brigadier general in April 1862, and Captain Thomas L. Casey was supervising coastal fortifications in Maine.

Now that he was a seafarer with a mechanical background, Weir enlisted in the US Navy at Boston on August 25, 1862, as an acting third assistant engineer, the equivalent of a midshipman. Before shipping out, he traveled to Cohoes and married Anna Chadwick on September 16. He soon left her and departed on the steamship *Rhode Island* to join the West Gulf Blockading Squadron, arriving on October 11 and being assigned to the USS *Richmond.*

The US Navy's role in the war was twofold. First, it formed a cordon along the Confederate coast to restrict commerce and seize the blockade runners that sought to deliver goods to Southern ports. At the same time, it conducted attacks to seize Southern ports and to take control of the Mississippi River. The USS *Richmond* was engaged in both of those efforts.

Commissioned as a *Hartford*-class steam sloop of war in 1860, the 225-foot USS *Richmond* carried both a square-rigged-ship rig and two direct-acting steam engines. The coal-fired engines generated 1,078 indicated horsepower to power a single screw, driving the ship at up to 9½ knots. Each engine, with a 58-inch cylinder and three-foot piston stroke, was fired by its own boiler. When Weir joined the ship, it carried 20 nine-inch Dahlgren guns in broadside and two rifled Parrott guns from the West Point Foundry at Cold Spring on pivots, one firing 80-pound shells and the other firing 30-pound shells. The crew totaled 269 men.

Although designed for sea service, the *Richmond* was now part of Admiral David Glasgow Farragut's squadron on the Mississippi River north of New Orleans. On March 14, 1863, Farragut attempted to run up the Mississippi past the Confederate stronghold at Port Hudson, Louisiana, to support federal military operations against Vicksburg, Mississippi. With the gunboat USS *Genesee* alongside, the *Richmond* followed Farragut's USS *Hartford* up the long reach of river below the bluffs at Port Hudson, receiving heavy fire from the Confederate forts.

"This was my first experience to be placed under fire, but with Farragut ahead and a silent prayer for protection we felt all would come out right," Weir recalled years later. As third assistant engineer, his position was alongside the ship's captain, Commander James Alden Jr., and the executive officer, Lieutenant Andrew B. Cummings, on the bridge over the deck amidships, where he conveyed orders to the engine room

On Thanksgiving Day, November 27, 1862, Weir wrote to his wife "Daisy," including this sketch of the Thanksgiving service on the deck of the USS *Richmond*. Looking aft toward the ship's wheel, he shows the officers seated to starboard and the crew gathered to port as Lieutenant Andrew Cummings stands at the flag-draped capstan to conduct the service. (Courtesy of the Pearce Museum, Navarro College, Corsicana, Texas)

through bell signals. During the bombardment, Lieutenant Cummings was mortally wounded when a shell took off his left leg as he stood beside Weir, and Weir himself was knocked down by splinters and a shell fragment that struck his belt buckle. Weir had just rung four bells to increase steam to round the bend in the river when a 42-pound Confederate shell pierced the hull and "a crash was heard in the Engine room, followed by the roar and shrieking hiss of escaping steam." The shell had struck the cast-iron steam chests of both boilers and knocked off the safety valves. Without power, and with the engine room and berthing room filled with scalding steam, the ship was towed downriver by the *Genesee*. Four of the ship's firemen would receive Medals of Honor for their bravery in securing the ship's engine room.[11]

For extended service on the Mississippi River, in June 1863, the ship's 80-pound Parrott rifle was replaced with a 100-pounder, and three additional cannons were added. At that time, Captain Thornton A. Jenkins replaced James Alden in command. On June 10, Confederate artillery moved close to shore and opened fire on the small mortar schooners that arced large shot from their "sconce kettles" into the Confederate works at Port Hudson. "We got underway, steaming up quietly, enjoying the exciting scene, and throwing a 100-pound shell from our pet Parrott as often as possible," wrote Weir to *Harper's Weekly*. "The rebels shot threw water up in fine style about our vessels. A few of their rifled shot came whizzing through our rigging. When just above the [ironclad USS] *Essex*, we let them have a broadside which knocked the dust about their ears in such a style that they concluded it best to close the action." The *Richmond* continued to support the siege of Port Hudson until its fall on July 9. With Admiral Farragut away, Captain Jenkins received the Confederate surrender on the *Richmond* that day. The surrender of Port Hudson, five days after the surrender of Vicksburg, opened the entire Mississippi River to US Navy control.[12]

On July 30, the *Richmond* left New Orleans and headed for the New York Navy Yard, where the vessel was repaired and overhauled. During the work, Weir may have had a chance to spend time with Anna. On October 12, the *Richmond* departed New York, calling at the navy bases at Port Royal, South Carolina, and Key West, Florida, before returning to New Orleans on November 1. In December, the vessel was dispatched to blockade the port of Mobile, Alabama. Again serving in the open waters of the Gulf of Mexico, the *Richmond*'s armament was cut back to 18 nine-inch Dahlgren guns in broadside and the pivoting 100-pound and 30-pound Parrott rifles in June 1864.

Weir was promoted to acting second assistant engineer on February 20, 1864. He wrote frequently to "Daisy," as he called Anna, solicitous of her health, frustrated by boredom and the ineptitude he saw in the navy, and happy when he was busy, even if it was maintaining the steam log by recording the small details of fuel consumption, steam pressure, and revolutions per minute of the engines. He might be tested on "Saturations

HARPER'S WEEKLY.

A JOURNAL OF CIVILIZATION.

VOL. VII.—No. 342.] NEW YORK, SATURDAY, JULY 18, 1863. [SINGLE COPIES SIX CENTS. $3.00 PER YEAR IN ADVANCE.

Entered according to Act of Congress, in the Year 1863, by Harper & Brothers, in the Clerk's Office of the District Court for the Southern District of New York.

THE BOMBARDMENT OF PORT HUDSON—THE 100-POUND PARROTT GUN OF THE "RICHMOND" AT WORK.—SKETCHED BY AN OFFICER OF THE NAVY.—[SEE PAGE 462.]

Facing page: Weir's first illustration in *Harper's Weekly*, "The Bombardment of Port Hudson—the 100-pound Parrott Gun of the 'Richmond' at Work—Sketched by an Officer of the Navy," appeared on the front page of the first issue to include news of the surrender of Port Hudson. "We are about two miles below the rebel batteries, which extend about three miles along the east bank of the river," wrote Weir on June 14, 1863. "With this gun we can reach their centre and most formidable works with ease, while with their 10-inch Columbiads they occasionally succeed in dashing the water up about us, few of their shots taking effect among our little fleet." (*Harper's Weekly*, July 18, 1863, G. W. Blunt White Library, Mystic Seaport Museum, Mystic, Connecticut)

"Bombardment of Port Hudson—A Mortar Schooner at Work—Sketched by a Naval Officer," was Weir's view of one of the six mortar schooners' "sconce kettles" firing a 13-inch, 200-pound explosive shell in an arc over the Confederate defenses on the 80-foot bluffs above the river. (*Harper's Weekly*, July 18, 1863, G. W. Blunt White Library, Mystic Seaport Museum, Mystic, Connecticut)

Officers pose at one of the USS *Richmond*'s "pet Parrotts," with the other one visible on the foredeck, ca. July 1863. At center is Captain Thornton A. Jenkins, who replaced Commander James Alden as commanding officer in June 1863 and received the surrender of Port Hudson on board the *Richmond* on July 9, 1863. He later commanded a division of the West Gulf Blockading Squadron from the *Richmond*. Acting Third Assistant Engineer Robert F. Weir may be one of the young men in the rear rank—possibly the one at center, whose high forehead and long nose resemble Weir family features. (2017.28, Dorothy Fairhurst Collection, Naval Historical Center, Washington, DC)

of different kinds of water in boilers—Expansion of steam—how much water & fuel is expended in generating such quantities of steam in different kinds of boilers." In winter, the engineering officers sometimes heated round shot in the boiler fireboxes to provide warmth in their quarters. Weir told Daisy, "The remembrance of your loveliness & purity is a sufficient spur to enable your husband to overcome many a tough obstacle."[13]

On August 5, 1864, Admiral Farragut got his fleet under steam to pass the forts defending Mobile Bay and seize the port. With the USS *Port Royal* alongside, the *Richmond* was in line behind the ironclad monitors and the USS *Brooklyn*. The monitor *Tecumseh* struck a "torpedo" or mine and sank immediately. When the *Brooklyn* then backed to avoid the torpedoes, the *Richmond* had to back, too, throwing off the line. "Damn the torpedoes . . . full speed ahead," ordered Admiral Farragut, and his USS *Hartford* steamed ahead past the *Richmond*. When the ships entered the bay, they engaged the Confederate ironclad ram *Tennessee* and three other vessels. The *Hartford, Richmond, Brooklyn, Lackawanna, Monongahela,* and three monitors attacked the *Tennessee*, battering it for an hour until it surrendered. The *Richmond* then joined the squadron in bombarding Fort Morgan. During this extended action, the *Richmond* received only slight damage. Twenty-nine members of the crew would receive Medals of Honor for bravery during the battle.[14]

Among Weir's sketches of on-board activities is this view of the engine-room crew cleaning one of the boilers of the USS *Richmond*. With an officer supervising, four sailors climb into the firebox to scrub out ash and clinkers. (Courtesy Mariners' Museum and Park, Newport News, Virginia)

Harper's Weekly published four of Weir's images of the Battle of Mobile Bay. Here, the 223-foot US Navy ironclad monitor *Tecumseh* capsizes and sinks after striking a Confederate "torpedo," or floating mine. The vessel sank in 30 seconds, taking down 94 of the 116 men on board, including Commander Tunis Craven. (*Harper's Weekly*, September 10, 1864, G. W. Blunt White Library, Mystic Seaport Museum, Mystic, Connecticut)

After the battle, the *Richmond* took station in Mobile Bay. Weir had been busy assessing the *Richmond*'s problematic engines and wrote and illustrated a report on them, which Assistant Secretary of the Navy Gustavus Fox ordered to be published. Weir made a point of sending a copy to his father. When he wrote to his young half-brother Charles Gouveneur Weir in March 1865, Weir described and illustrated an "Attack of (pseudo) Rebel Torpedoboat upon U.S.S. Richmond" a week earlier. A drifting log had approached the *Richmond* one night, and the ship's sentries had feared an attack and fired on it. "I hope you are getting along well in your studies—and have plenty of exercise and fun to back it up with," he told Charlie. "I hope I can have some time to myself this summer—when if we are together I propose making some interesting experiments with you boys," signing off "Keep up merry hearts—the last ditch is in sight & oil is up."[15]

During his service on the *Richmond*, off watch on deck, or seated at the table in the engineers' six-man bunk- and mess-room, Weir continued sketching in pen-and-ink and, sometimes, watercolor. He depicted details of life on board and notable incidents, sometimes contributing sketches to *Harper's Weekly*. On July 18, 1863, the magazine published two of his sketches of the *Richmond*'s forward Parrott rifle and a mortar schooner firing on Port Hudson, plus the letter in which he described the action. *Harper's Weekly* published four of his drawings, including a bird's-eye view of the *Richmond*, on February 13, 1864. In the September 10, 1864, issue, he was named as the artist of four images of the Battle of Mobile Bay, and

Titled "Farragut's Victory in Mobile Bay—The Capture of the Rebel Ram Tennessee," Weir's sketch shows the Confederate ironclad ram *Tennessee* surrounded by the US sloops of war *Brooklyn*, *Lackawanna*, and *Monongahela* and the monitors *Manhattan* (right), *Chickasaw* (astern), and *Winnebago* (left). (*Harper's Weekly*, September 10, 1864, G. W. Blunt White Library, Mystic Seaport Museum, Mystic, Connecticut)

Weir's aerial view of the USS *Richmond* in heavy seas was possibly influenced by concepts in aerial perspective learned from his father. The wood engravers at *Harper's* simply copied Weir's drawing, producing a mirror image of the original when the wood block was printed. This is one of four images by Weir in the issue. (*Harper's Weekly*, February 13, 1864, G. W. Blunt White Library, Mystic Seaport Museum, Mystic, Connecticut)

Harper's Weekly published other illustrations under his name in September and October 1864 and May 1865.

The *Richmond* continued to serve in Mobile Bay and off Pensacola, Florida, until April 1865. Early that month, the Confederate capital of Richmond, Virginia, fell and the Confederate government fled. The Confederate Army of Northern Virginia surrendered on April 9, and gradually, hostilities ceased. The *Richmond* headed to the mouth of the Mississippi, arriving on April 23. A few days later, it intercepted the Confederate ram *Webb* running for the sea. Cut off by the *Richmond*, the *Webb* was destroyed by its crew to avoid capture (*Harper's Weekly* would publish Weir's sketch of the incident on May 20, 1865). On May 26, the

Confederate Trans-Mississippi Department surrendered, and by June, the war was over. The *Richmond* returned to the Atlantic and arrived at the Boston Navy Yard on July 10, 1865. Weir was discharged from the vessel on July 11 and resigned his naval commission the next day. Two days later, with the rest of the crew discharged or reassigned, the USS *Richmond* was decommissioned.

After the war, Weir retained a relationship with *Harper's Weekly*, and on December 22, 1866, it published his "Taking a Whale," an image similar to his journal rendering of "Taber Tom." This time, he had apparently been asked to illustrate an incident in a whaling story written by Augustus Comstock, a sometime whaleman and popular "dime-novel" author who wrote under the pen name "Roger Starbuck."[16]

"Taking a Whale.—Sketched by R. Weir" is similar to Weir's sketch of "Taber Tom" in his journal, and to his painting. For this full-page *Harper's Weekly* woodcut, he relied on his experiences to illustrate a dramatic moment in a story by "Roger Starbuck"—actually Augustus Comstock, a sometime whaleman and author of seafaring and western adventure "dime-novels." (*Harper's Weekly*, December 22, 1866, G. W. Blunt White Library, Mystic Seaport Museum, Mystic, Connecticut)

In a composition very similar to his "Taber Tom" journal sketch, Weir painted *Taking a Whale / Shooting a Whale with a Shoulder Gun*, a large watercolor that perhaps represents Second Mate Welch firing his bomb-lance gun. In each image, the boats direct the eye to the action, but in the painting, the whale exhales in a white spout (properly angled forward for a sperm whale), while in the journal sketch, the lance has pierced the lungs and blood can be seen in the spout. (Courtesy of the New Bedford Whaling Museum)

According to the federal census, in 1870, Weir was employed as a civil engineer and living with Anna in a New York boardinghouse alongside several clerks and lawyers. His personal estate was valued at $5,000 (about $117,400 in 2023). He had an engineering or design position with the construction department of the Croton Water Works, which supplied drinking water to New York City through the 40-mile-long Croton Aqueduct and a series of valves and gates to maintain the flow to reservoirs in Manhattan. His half-brother Charles, or Charlie, also had an affinity for engineering, serving as a civil engineer in New York City for improvements to the East River.

By 1875, Robert and Anna had moved north to Cohoes to live with her elderly parents. Robert took a position in the family's Ontario Knitting Company, and after William Chadwick died in 1880, Robert was termed a "manufacturer" in the federal census, living in Cohoes with Anna, possibly his half-sister Annie, and a female servant.

During Robert's absence from New York City, his father retired from his position at West Point in 1876 and moved to Hoboken, New Jersey, across the Hudson from his sons in New York City. He continued to paint until his death in 1889.

Art flourished in several generations of the Weir family. Robert's younger brother John Ferguson Weir studied under their father and, at 21, obtained a commission for a painting of the Hudson from West Point. It was so well received that he quickly established himself as one of New York's talented young artists. Elected to the National Academy of Design, like his father, John then spent a year studying art in Europe before becoming the first director of the School of Fine Arts at Yale in 1869, continuing as director or dean until 1913. While his large post–Civil War paintings of the West Point Foundry reinforced his early acclaim, his output declined during his years as a teacher and, much like his father, his style changed little. In 1866, John married Hannah French, daughter of the West Point chaplain, Rev. John W. French, and their daughter Edith (1875–1955) would become a painter of miniatures. The eldest brother Walter's daughter Irene (1862–1944) became an artist and art teacher. After brother-in-law Truman Seymour retired from the US Army in 1876, he and Louisa moved to Europe, where both of them sketched and painted.

While setting up the Yale School of Fine Arts, John requested the assistance of his half-brother Julian Alden Weir, who was then a 21-year-old student at the Ecolé des Beaux-Artes in Paris. While John emphasized teaching, Julian became a highly regarded artist and traveled often to Europe, where he witnessed the forefront of stylistic changes. While at first alarmed at the move toward impressionism, he would become one of America's foremost impressionists.

The Weir siblings also faced tragedy. In 1879, Robert's half-brother William Bayard Weir, an 1870 graduate of West Point, was killed while on duty in Colorado. By that time, the eldest brother, Walter, had been committed to an asylum for mental instability. Since taking a commission in the 5th US Artillery in 1861, brother Gulian Verplanck Weir had been considered a brave and skilled artilleryman, remaining in the service after the war. However, he apparently felt that he had failed terribly by having three guns captured during the Battle of Gettysburg in 1863. Despite having five young children, Gulian Weir committed suicide at Fort Hamilton, Brooklyn, in 1886.[17]

Robert and Anna Weir had no children. They moved to Montclair, New Jersey, about 1892, and around that time, Weir joined the Empire City Subway Company, which was established in 1891 to manage the underground conduits for electrical and other uses in Manhattan and the Bronx. From the company's office on West 38th Street in New York City, Weir would finish his career designing machinery and calculating ways to run and seal underground lines. Also in the 1890s, he wrote down some of his Civil War naval remembrances and composed stories, none of which seem to have been published.[18]

Robert Fulton Weir died at age 69 on January 17, 1905, at his home in Montclair. Anna moved back to Cohoes and died in 1910. They are buried with Anna's family at the Oakwood Cemetery in Troy, New York.

Much of Robert F. Weir's artwork survives. His *Clara Bell* journal and a few illustrated letters, recollections, stories, and drawings are preserved in the G. W. Blunt White Library at Mystic Seaport Museum in Mystic, Connecticut. Fifty-three of his naval sketches are at the Mariners' Museum and Park in Newport News, Virginia, and 30 of his illustrated letters to "Daisy" are in the Pearce Civil War Collection at the Pearce Museum at Navarro College in Corsicana, Texas. The New Bedford Whaling Museum in New Bedford, Massachusetts, holds four paintings, including the large watercolor *Taking a Whale/Shooting a Whale with a Shoulder Gun*, and an illustrated book. Wood-engraved renderings of about 15 of his illustrations can be found in *Harper's Weekly*.

An unsettled son in a large, prominent, and talented family, Robert Fulton Weir ran off to sea to escape an imagined embarrassment to them, proved himself as a bold whaleman while enhancing his observational ability and his artistry, then served his country bravely in the US Navy during the Civil War before taking up a precise and demanding position as a civil engineer in New York City, assisting his father-in-law in managing a textile mill, and continuing his creative efforts by writing stories, real and imaginary. He hinted at his own life when he quoted his character Buttons, the elevator operator in one of his stories: "My life here has made me realize that no matter how humble one may be, or how obscure one's profession, he often has it in his power to do some good in the world, and unconsciously to help others by sympathy, [respect], or example."[19]

Notes

1. Alexander Starbuck, *History of the American Whale Fishery* (1876; reprint, Secaucus, NJ: Castle Books, 1989).
2. "Copy of Expenses on Shipboard Drawn from the Ships Stores Slop Shop," in Robert F. Weir, *Clara Bell*, Log 164, G. W. Blunt White Library, Mystic Seaport Museum, Mystic, CT.
3. Charles H. Robbins, *The Gam: Being a Group of Whaling Stories* (1899; Salem, MA: Newcomb & Gauss, 1913).
4. Whaling Crew List Database, New Bedford Whaling Museum, https://www.whalingmuseum.org/online_exhibits/crewlist/search.php, accessed September–November 2023.
5. Anthony J. Connors, *Went to the Devil: A Yankee Whaler in the Slave Trade* (Amherst, MA: University of Massachusetts Press, 2019), 81–91.
6. Whaling Crew List Database, New Bedford Whaling Museum, https://www.whalingmuseum.org/online_exhibits/crewlist/search.php, accessed September–November 2023.
7. Connors, *Went to the Devil*, 95–96.
8. Ibid., 96–101. The Whaling Crew List Database identifies Boatsteerer William Robinson as a 23-year-old from Providence, Rhode Island.
9. Ibid., 101–2.
10. Ibid., 102–4.
11. Robert F. Weir, "Some War Remembrances, Sept. 1897," Folder 1, Collection 245, G. W. Blunt White Library, Mystic Seaport Museum, Mystic, CT.
12. *Harper's Weekly*, July 18, 1863, 449, 452, 462.
13. Letters from Weir to "Daisy," in the Robert F. Weir Papers, Pearce Civil War Collection, The Pearce Museum, Navarro College, Corsicana, Texas, quoted in Sheritta Bitikofer, "Sailor and an Artist—Robert Weir of the USS *Richmond*," *Emerging Civil War*, March 10, 2022, https://emergingcivilwar.com/2022/03/10/sailor-and-an-artist-robert-weir-of-the-uss-richmond/, accessed October 2023.
14. "Robert F. Weir, "Recollections II, Mobile Bay," Folder 2, Collection 245, G. W. Blunt White Library, Mystic Seaport Museum, Mystic, CT.
15. Weir to "Charlie," March 28, 1865, VFM 1298, Manuscripts Collection, G. W. Blunt White Library, Mystic Seaport Museum, Mystic, CT. Weir's published article on the USS *Richmond*'s engines has not been located. The engines would be replaced in 1866.
16. *Harper's Weekly*, December 22, 1866, 607, 613.

17. For the later activities of the Weir family members, see Marian Wardle, ed., *The Weir Family, 1820–1920: Expanding the Traditions of American Art* (Provo, UT and Hanover, MA: Brigham Young University Museum of Art and University Press of New England, 2011), esp. 157–70.
18. Robert F. Weir, Sketches and Mechanical Drawings and Miscellaneous, Folder 6, 7, Collection 245, G. W. Blunt White Library, Mystic Seaport Museum, Mystic, CT.
19. Robert F. Weir, Buttons Story, Folder 4, Collection 245, G. W. Blunt White Library, Mystic Seaport Museum, Mystic, CT.

Bibliography

"The Adventures of a Haunted Whaling Man," *American Heritage* 28, issue 5 (August 1977).

Bitikofer, Sheritta. "Sailor and an Artist—Robert Weir of the USS *Richmond*." *Emerging Civil War*, March 10, 2022. https://emergingcivilwar.com/2022/03/10/sailor-and-an-artist-robert-weir-of-the-uss-richmond/.

Browne, J. Ross. *Etchings of a Whaling Cruise*. New York: Harper & Brothers, 1846.

Browning, Robert M. "Reflections of the Civil War at Sea: Robert Fulton Weir used his artistic talents to sketch scenes of shipboard life during the sectional conflict." *Naval History Magazine*, April 2020. https://www.usni.org/magazines/naval-history-magazine/2020/april/reflections-civil-war-sea.

Busch, Briton Cooper. *Whaling Will Never Do for Me: The American Whaleman in the Nineteenth Century*. Lexington, KY: University Press of Kentucky, 1994.

"Chadwick Family Papers." New York State Library, Albany, NY. https://www.nysl.nysed.gov/msscfa/sc16555.htm.

Connors, Anthony J. *Went to the Devil: A Yankee Whaler in the Slave Trade*. Amherst, MA: University of Massachusetts Press, 2019.

Creighton, Margaret S. *Rites and Passages: The Experience of American Whaling, 1830–1870*. Cambridge and New York: Cambridge University Press, 1995.

Dana, Richard Henry. *Two Years Before the Mast*. New York: Harper & Brothers, 1840.

Davis, Lance E., Robert E. Gallman, and Karin Gleiter. *In Pursuit of Leviathan: Technology, Institutions, Productivity, and Profits in American Whaling, 1816–1906*. Chicago: University of Chicago Press, 1997.

Delano, Reuben. *Wanderings and Adventures of Reuben Delano, Being a Narrative of Twelve Years Life in a Whale Ship*. Worcester, MA: Thomas Drew Jr., 1846.

Dewey, Charles P. Journal of a whaling voyage on the barks *John Dawson* and *Clara Bell*, 1855–1858. Log 590a, New Bedford Whaling Museum, New Bedford, MA. It includes two pencil sketches by Weir.

Dolin, Eric Jay. *Leviathan: The History of Whaling in America*. New York: W. W. Norton, 2007.

Druett, Joan, ed. *"She Was a Sister Sailor": The Whaling Journals of Mary Brewster, 1845–1851*. Mystic, CT: Mystic Seaport Museum, 1992.

Dyer, Michael P. *"O'er the Wide and Tractless Sea": Original Art of the Yankee Whale Hunt*. New Bedford, MA: New Bedford Whaling Museum, 2017.

Ellis, Richard. *The Great Sperm Whale: A Natural History of the Ocean's Most Magnificent and Mysterious Creature*. Lawrence, KS: University Press of Kansas, 2011.

Harper's Weekly, July 18, 1863; February 13, August 20, September 3, September 10, September 24, October 22, 1864; May 20, 1865; December 22, 1866.

"History of West Point Foundry." https://web.archive.org/web/20070629101949/http://www.scenichudson.org/land_pres/wpfp_research.htm.

Hohman, Elmo P. *The American Whaleman: A Study of Life and Labor in the Whaling Industry*. London: Longmans, Green & Company, 1928.

Holy Innocents Episcopal Church. "History." https://holyinnocents2.wixsite.com/hihf/history.

Lytle, Thomas G. *Whalecraft*. Website, www.whalesite.org.

McQuarrie, Gary. "Robert Fulton Weir: Sailor-Artist for Harper's." *Civil War Navy—The Magazine*, Spring 2020, 43–49. https://civilwarnavy.com/wp-content/uploads/2020/04/Robert-Fulton-Weir-Sailor-Artist-for-Harpers.pdf.

Melville, Herman. *Moby-Dick*, 1851; reprint, Evanston, IL, and Chicago: Northwestern University Press and The Newberry Library, 1988.

New Bedford Directory. New Bedford, MA: Charles Taber & Company, 1856.

New Bedford Whaling Museum. "Whaling Crew List Database." https://www.whalingmuseum.org/online_exhibits/crewlist/search.php?.

Nordhoff, Charles. *Whaling and Fishing*. New York: Dodd, Mead & Company, 1855.

Olmsted, Francis Allyn. *Incidents of a Whaling Voyage*. New York: D. Appleton & Company, 1841.

Robbins, Charles H. *The Gam: Being a Group of Whaling Stories*. 1899; Salem, MA: Newcomb & Gauss, 1913.

"Robert Weir Dead," *New York Times*, January 18, 1905.

Starbuck, Alexander. *History of the American Whale Fishery*. 1876; reprint, Secaucus, NJ: Castle Books, 1989.

"Wallace, William" [Weir, Robert F.]. Receipt, August 18, 1855, VFM 1743, Manuscripts Collection, G. W. Blunt White Library, Mystic Seaport Museum, Inc., Mystic, CT.

Wardle, Marian, ed. *The Weir Family, 1820–1920: Expanding the Traditions of American Art*. Provo, UT: Brigham Young University Museum of Art; Hanover, MA: University Press of New England, 2011.

Weir, Robert F. Journal, Bark *Clara Bell*, 1855–1858, Log 164, Manuscripts Collection, G. W. Blunt White Library, Mystic Seaport Museum, Mystic, CT.

Weir, Robert F. Letter to Charles G. Weir, March 28, 1865, VFM 1298, Manuscripts Collection, G. W. Blunt White Library, Mystic Seaport Museum, Mystic, CT.

Weir, Robert F. Letters to John F. Weir, 1869, John Ferguson Weir Papers, MS 550, Archives at Yale, Yale University, New Haven, CT.

Weir, Robert F. Paintings and illustrated book, New Bedford Whaling Museum, New Bedford, MA.

Weir, Robert F. Papers, Collection 245, Manuscripts Collection, G. W. Blunt White Library, Mystic Seaport Museum, Mystic, CT.

Weir, Robert F. Papers, MS0007, Mariners' Museum Library, Mariners' Museum and Park, Newport News, VA.

Weir, Robert F. Papers, Pearce Civil War Collection, The Pearce Museum, Navarro College, Corsicana, TX.

Whitecar, William B. *Four Years Aboard the Whaleship. Embracing Cruises in the Pacific, Atlantic, Indian, and Antarctic Oceans. In the Years 1855, '6, '7, '8, '9*. Philadelphia: J. B. Lippincott, 1860.

A Glossary of Whaling Terms

Ambergris. A waxy material produced in the digestive system of sperm whales in response to irritation (such as from the beaks of the giant squid they eat); ambergris is sometimes found floating at sea. Because it was an ideal stabilizer for scents in perfume, it became a very valuable occasional product for whaleships.

Bark. A sailing vessel with three or more masts, square-rigged on all but the aftermost mast, which has only fore-and-aft sails. All whaling barks were three-masted.

Barrel. A wooden cask holding about thirty-two gallons of liquid. Whalemen calculated quantities of whale oil taken in terms of 31.5-gallon barrels.

Bearers. Timbers fastened to the side of the vessel and standing upright, two between each set of davits, to steady the boats resting on the cranes, which pivot from the bearers.

Bends. A run of several thick planks along the widest part of the hull, between waterline and the planksheer at the deck. The bends strengthened the hull and absorbed chaffing.

Bible Leaves. The whalemen's term for the minced horse pieces of blubber, with slices cut through the fatty side, that were ready for rendering in the tryworks.

Bitter End. The inboard end of any line.

Blackfish. The whalemen's term for the pilot whale, a species of small, toothed whale. Pods of pilot whales were often encountered in the North Atlantic between New Bedford and the Azores, giving whaleboat crews a chance to practice chasing, harpooning, and lancing whales.

Blanket Piece. The strip of blubber, about six feet wide, that was cut away from the whale in a spiral pattern during the cutting-in process.

Blubber. The fatty substances that encase and insulate the body of a whale. Blubber ranges from several inches to more than a foot thick, depending upon the size and species of whale. When rendered in a whaleship's try-pots, the oil separated from the connective tissue, much as if one were frying salt pork or bacon. The process was called "trying out," or more commonly among whalemen, "boiling."

Blubber Room. The below-deck space between the crew's forecastle and the boatsteerers' and idlers' bunkrooms. This long, low room was used for cutting the large strips of blubber into smaller "horse" pieces.

Boatheader. The commander of a whaleboat. Normally, the captain headed the starboard boat, the first mate headed the larboard boat, the second mate headed the waist boat, and the third mate headed the larboard bow boat (with a fourth mate heading the starboard bow boat if carried). The boatheader steered the whaleboat during the chase and then went forward to kill the whale with a lance.

Boatsteerer. Also known as the harpooner, he pulled the forward oar in a whaleboat and had the responsibility of harpooning the whale. Once that job was done, he went aft to steer while the boatheader killed the whale with a lance.

Bomb Lance. A hollow metal projectile filled with gunpowder that was fired from a shoulder gun into a whale to explode and kill the whale. Introduced around 1850, the bomb lance was especially useful in hunting aggressive whales or bowhead whales, which might escape under the ice before they could be lanced by hand.

Box. A portion of the forward deck of a whaleboat, margined at the after end by the "clumsy cleat."

Box Warp. Sometimes called the stray line, this part of the whale line was attached to the harpoon and coiled down in the "box" in the bow of the boat before passing out through the notched stem. The box warp gave the boatsteerer some slack line as he darted his harpoon.

Bulwarks. The high, strong rail around a ship's deck.

Case. The whalemen's term for the compartment in the sperm whale's forehead that contains waxy spermaceti and may be involved in echolocation.

Chains. Thick, narrow platforms outboard of each mast to widen the angle of support for the masts provided by the shrouds. Each shroud was tightened with lanyards laced through a pair of deadeyes, one spliced to the shroud and the other anchored to the chains.

Clew. The lower corner of a square sail.

Clumsy Cleat. A heavy wooden brace that extends across the after end of the foredeck or box in the bow of a whaleboat. It has a notch used by the harpooner to brace his thigh as he darted his harpoon. Projecting up through it is the rod to take the forward hoisting ring or shackle. The kicking strap on its surface is knotted underneath.

Course. The lower square sail on each mast. Spoken of individually, the sail was often called by the name of the mast from which it was set.

Cracklings. The whalemen's term for the flesh and connective tissue that remained after the oil was rendered out of the blubber. Cracklings were used to fuel the tryworks fires, producing oily, black smoke.

Cranes. Hinged, triangular wooden brackets that swing into a horizontal position from the bearers. When a whaleboat was hoisted in the davits from the water, it was set down on the cranes.

Crosstrees. The pairs of short, horizonal spars at the points where the topmast and topgallant mast overlap and where the topgallant and royal masts overlap. The crosstrees spread the shrouds supporting the topgallant and royal mast sections and can be used as lookout platforms.

Crotch. A light, forked timber on the starboard gunwale of a whaleboat that served as a rest for the two working harpoons until they were darted.

Cutting In. The process of cutting the blubber away from the carcass of the whale as it lay in the water alongside the ship. The blubber was sectioned in a wide, spiraling blanket piece that was cut away from the musculature underneath. The vessel's mates usually performed the operation from a cutting stage rigged out over the starboard side.

Cutting Spade. A tool shaped like a spade, eight or ten inches across and sharpened along the bottom edge for cutting blubber from the whale after it was alongside the ship.

Cutting Stage. The platform suspended along the starboard side by the gangway, on which the mates stood while wielding their cutting spades during the cutting in process. In the 1850s, the cutting stage was two small platforms, one on either side of the gangway. Later, the cutting stage was several long planks extending parallel to the starboard side of the ship for about 12 feet or more, with iron stanchions holding a pole railing along the length of the staging to lean over while cutting.

Cutting Tackle. Sets of double-block tackles that were hooked or shackled into a chain or rope "necklace" around the mainmast head and used for hoisting in the blubber.

Dart. The whalemen's term for the act of harpooning a whale. Harpoons could be thrown, but the preferred method was to bring the whaleboat right up to the whale so the harpoons could be driven through the blubber to anchor in the whale's flesh.

Davits. Heavy wooden timbers fastened to the sides of a whaler, two for each whaleboat; these were the unique identifying mark of whaling ships. The tops were steam-bent in a curve to form a projection. The top was slotted and fitted with sheaves or pulley wheels through which rope falls were led to form tackles for hoisting the whaleboats. The *Clara Bell* had three pairs to larboard and one to starboard as a four-boat whaler.

Drawed. The whalemen's ungrammatical synonym for "drew." After a whale was harpooned, the harpoon sometimes pulled out and the whaleman's description was that he had "drawed his iron."

Drogue. Usually a wooden platform about 2 to 2.5 feet square, bridled from each corner with whale line, or sometimes a wooden bucket. If a whale ran out all of the line in the tubs, the drogue was made fast to the bitter end of the second line and tossed overboard to impede the whale's progress.

Fathom. A nautical measure. One fathom equals six feet.

Finback (Fin) Whale. A species of rorqual whale that lives along the coasts and was sometimes taken by small-boat whalemen. Although common along the New England shoreline, the species was never important to the New England whaling industry.

Fluke Chain. A heavy chain that was shackled around a dead whale just forward of the flukes to secure it alongside the ship, tail forward, for cutting in.

Fluke Pipe. A hawse hole, cut low through the starboard bulwarks, through which the fluke chain was led to be attached to the fluke bitt (post) abreast of the foremast.

Footropes. Ropes strung horizontally below a ship's yards on which the crew would stand while setting or furling the square sails.

Galley or Gallied. To startle a whale and cause it to swim away before it can be harpooned. The galley is also the deck structure where all cooking is done.

Gamming. When two whaleships met at sea, if the weather was favorable, they hove to and one captain took his boat to the other vessel, while the other vessel's first mate took his boat's crew to the first vessel. Gamming was very common when a captain's wife was aboard.

Goose Pen. A watertight, boxlike structure built on deck on which the brick tryworks was set. It was kept filled with water whenever there was a fire in the tryworks to prevent the deck from being scorched or set afire.

Grounds. Areas in seas and oceans where whales were found to be plentiful. For the *Clara Bell*, those were the waters around the Azores, between Africa and St. Helena, and around the island of Madagascar.

Gunwale. The rail or edge of an open boat.

Harpooner. Synonymous with boatsteerer, which was the more common term in American whaleships.

Hawse Hole. An oval-shaped cut through the bulwarks of a vessel into which an iron casting is inserted and through which lines or chains are passed.

Heave Short. To take in the anchor chain by operating the windlass until the chain is vertical and the anchor is about to lift off the seabed.

Heave To (Hove To). To hold a position at sea, a ship's crew would brace either the fore or main topsail aback, with the wind on the forward side to stop the vessel, while the other topsail remained full to counteract the backward force. To heave to in very heavy weather, the crew would set the wheel to steer toward the wind while setting just enough sail to push the bow off the wind, thus balancing the action of the rudder so the ship remained in a safe orientation to ride over the waves.

Horse Piece. The whalemen's term for the segments cut from the blanket piece of blubber in the blubber room. Horse pieces were about six feet long and a foot wide. They were then brought on deck to be minced into bible leaves.

Humpback Whale. A species of rorqual whale that had populations in the North and South Atlantic, around Madagascar in the Indian Ocean, and off the west coast of Central America and lower California in the Pacific. Humpbacks were fast and their carcasses often sank when killed, so they were not a major prey for New England whalers.

Iron. The whalemen's name for a harpoon.

Kicking Strap. A heavy piece of line made fast at each end through holes in the "clumsy cleat" and under which the whale line passed before being coiled down in the box. The kicking strap helped control the whale line after the whale was harpooned and began to tow the whaleboat.

Knot. A measure of a ship's speed equaling one nautical mile per hour (1.15 land miles per hour). Speed was calculated with a logline, with knots spaced along it, that was streamed astern for a number of seconds (measured by a sand glass). The number of knots that ran out in that time gave the number of nautical miles the vessel would travel in an hour. See **Logline**.

Lantern Keg. A conical-shaped cask with flat ends that was filled with a lantern and some hardtack crackers and stowed in each whaleboat in the event that the boat went astray and needed some sustenance and a light to signal its vessel.

Lapstrake. A form of small-boat construction where the edges of the planks are lapped one over the other and fastened at close intervals through the edges.

Larboard. A nautical term for left, especially the left side of a vessel. In the 1800s, other sea services changed to the use of "port" for left, but whalemen continued to use larboard.

Lay. Rather than receiving a set wage, whalemen were paid in the form of a share (lay) in the proceeds of the catch, less certain expenses. Lays varied from 1/15th to 1/200th, depending on the rank or experience of the men who received them, and were agreed upon as a condition of signing aboard before the voyage began.

Loggerhead. A stout hardwood post anchored in the afterdeck of a whaleboat, around which the whale line was looped to serve as a brake on the line as a harpooned whale tried to escape. The crew kept the line and loggerhead doused with water to prevent them from igniting through friction.

Logline. A knotted line wound on a reel, with a triangular wooden "chip" attached to the end. To determine a vessel's speed, the line was streamed, with the chip anchoring the end in the water. As the sand in a 14- or 28-second sand glass ran out, the line unreeled, with knots spaced at either 42- or 48-foot intervals. The number of knots that ran out during the time indicated the number of nautical miles (1.15 land miles) per hour the vessel was traveling.

Luff. To bring a sailing vessel up, bow into the wind, with sails shaking so that the vessel loses headway.

Nantucket Sleigh Ride. The whalemen's term for when a harpooned whale swam on the surface, towing the whaleboat at a high speed.

Preventer. A term for extra, a backup, or a temporary reinforcement. An extra boatsteerer was called a preventer boatsteerer, and an extra line added as a reinforcement might be called a preventer guy, preventer sheet, or preventer stay.

Right Whale. This baleen whale with thick blubber was the original species hunted in the Atlantic and remained one of the principal whales hunted by New England whalemen, who tracked them in the South Atlantic and throughout the Pacific and Indian oceans. By the 1850s, it was valued for its baleen as well as for its oil.

Rorquals. The largest group of baleen whales, including the finback, gray, humpback, and sulphur-bottom (blue) species. They are distinguished by "pleated throat grooves" that allow them to stretch their mouths greatly when feeding. Fast-swimming and prone to sinking when killed, rorquals were not widely hunted by New England whalers.

Royal. The sail above the topgallant sail, and usually the uppermost sail on any whaleship's mast.

Scrimshaw. A folk art practiced primarily by whalemen who used sperm-whale teeth, baleen (which they called whalebone), and jawbone (which they called panbone) to engrave pictures upon, or to make pie crimpers, corset busks, toys, and other articles that they took home to their wives or sweethearts. Synonyms were scrimshon and scrimshander as well as other variations of the word.

Ship. As a specific designation of rig, it means a square-rigged sailing vessel with three or more masts, all square-rigged. (Whaleships were never more than three-masted.) In generic terms, it can be used to speak of any large vessel.

Shook. The disassembled components of a cask or barrel, including staves, heads, and hoops. To save space, whaleships carried many barrels as shooks, which were reassembled by the ship's cooper as needed.

Shrouds. The term for the stays that supported the masts laterally. The tension of the stays was controlled by tightening the lanyards that ran through a pair of wooden deadeyes at the base of each stay. The shrouds were laced with horizontal ratlines to become rope ladders for the crew to go aloft.

Slide Boards. Light springy boards, about eight to ten inches wide, bent and fastened vertically to the side of the ship between the bearers or davits to hold the side of the whaleboat away and keep it from catching the gunwale on any projections as it was lowered and raised.

Slop Chest. A supply of tobacco, clothes, sheath knives, and sundries carried on board ship for the crew to purchase. The price was charged against their lay, with interest, and settled up when they were paid off at the end of the voyage. The captain usually reaped the profits.

Sound. When a whale was harpooned, it often dove and headed for the bottom. This was called sounding.

Sperm Whale. A species of toothed whale that was generally found in warmer equatorial waters but also ranged widely into both the North and South Atlantic and the Pacific. Because of the fine quality of its oil and the waxy spermaceti in its forehead "case," the sperm whale became the most desirable species for New England whalemen to hunt from the 1750s to the 1850s.

Starboard. The nautical term for right, especially the right side of a vessel. The term apparently derives from Norse-period vessels that were steered with an oar on the right, or steering side.

Stay. One of the fixed pieces of rigging that supports the masts. Originally fashioned from heavy hemp rope, after the 1860s stays were commonly made of strong but thinner wire rope. On board, stays were named for the portion of masts that they supported.

Stove. Smashed, usually by a whale. A whale would often thrash around after being harpooned or lanced and in doing so might damage or destroy the whaleboat, which would be referred to as a stove boat.

Studding Sail. A stun'sl or stud'sail to seamen, this lightweight sail was set outboard of a squaresail with lightweight booms at head and foot that extended outboard of the yardarms through rings. Used to increase a vessel's sail area when sailing downwind, stun'sls could be set on one or both sides of the squaresails and in varying combinations from mast to mast.

Tack. A word with several nautical meanings. As a noun, it means the line attached to the windward clew of the courses and also to the orientation to the wind in which a vessel is sailing: on the starboard tack when the wind comes from the right side and on the larboard tack when the wind comes from the left side. As a verb, it means to change that orientation by steering up and swinging the bow through the wind, bracing the sails around to take up the new tack.

Thole Pins. Stout wooden pegs inserted in a whaleboat's rail in pairs to act as oarlocks for securing the oars while rowing. Thole pins were later replaced by more durable iron oarlocks.

Tonnage. A figure arrived at by computing a vessel's length, beam (breadth), and depth (calculated as half the breadth) in a somewhat arbitrary formula, to be used in registering or documenting a vessel. It suggests the cargo volume of the hull but is not a measure of vessel weight or displacement.

Topgallant Mast. The third section of mast above the deck from which was set not only the topgallant sail but also the royal above it.

Topsail. The square sail set above the course on the section of mast above the lower mast. Beginning in the 1850s, the topsail was frequently divided into two sails with an extra yard or spar and became double topsails, which were easier to furl and manage in heavy winds.

Tryworks. A furnace built of bricks around two large cast-iron kettles, or try-pots, in which the blubber was rendered—tried out—to produce oil. It was positioned over the water-filled goose pen on deck aft of the fore hatch.

Waif. The whalemen's term for a small flag carried in a whaleboat that could be planted on a dead whale to claim it when several vessels were whaling in proximity.

Wear. The opposite of the verb tack, to change a sailing vessel's direction and orientation to the wind by steering away and swinging the stern through the wind.

Whale Line. Often called a tow line by whalemen, this was the line secured to the harpoon by which a whaleboat attached itself to a whale. Whale line was usually ¾-inch long-fibered manila of the best grade. Whaleboats normally carried 1,800 feet of whale line, divided into two tubs. When breaking out a new coil, the line was led up through a temporary block aloft and down to the boatsteerer on deck to be carefully coiled into the line tub to prevent possible kinking.

Whalebone. The whalemen's term for baleen, which is actually keratin, the component of human fingernails, not bone.

Windlass. A large, horizontal, revolving timber near the bow for hoisting anchors, blanket pieces of blubber, and other heavy objects. The *Clara Bell* had a pump-break windlass, with a ratchet mechanism operated by two long iron windlass bars pumped up and down by the crew to turn the windlass. As the windlass turned, a wooden pawl dropped into notches in the center of the windlass barrel to keep it from turning backwards.

Yard. A horizontal spar to which was fastened the head or upper edge of a square sail.

Yardarm. The end of a yard.